JFE
20- 5102

"Firmly situated in the discipline of history, but drawing upon a wide variety of theories, methodologies, and case studies from memory studies, transitional justice, and other interdisciplinary fields, Jelena Đureinović's timely book is an excellent and ground-breaking study into the problematic issue of memory politics in contemporary Serbia and its ramifications for other Yugoslav successor states. Đureinović's cutting-edge research will be eye-opening for scholars working not only on the Balkans but for those outside the region, revealing how post-conflict and post-communist societies like Serbia are susceptible to manipulation by politicized mnemonic actors."

Vjeran Pavlaković, University of Rijeka, Croatia

"In this well-researched and very convincing book, Jelena Đureinović demonstrates the processes of systematic politics of right-wing revisionism in Serbian history politics, adding new insights to our understanding of strategies, agency and power within history and memory politics. Đureinović's thought-provoking work draws our attention to the dangers of forgetting, deliberately ignoring and downplaying crimes of the past, and to the dynamics of reinterpreting history to fit political demands and needs in the present. We should remember that such revisionism inevitably contributes to changing society's understanding of the present and thereby also to shaping fears and expectations of the future, and Đureinović's excellent research is an important reminder about how this works."

Tea Sindbæk Andersen, University of Copenhagen, Denmark

"This outstanding book is the first comprehensive look at the remarkable transformation of political memory of World War II in contemporary Serbia. Đureinović convincingly demonstrates that in its commitment to anticommunism, Serbia has embarked on a full-scale revision of its WWII memory. This study is a timely warning of the seriously political consequences of playing politics with the past."

Jelena Subotić, Georgia State University, USA

The Politics of Memory of the Second World War in Contemporary Serbia

Exploring the concepts of collaboration, resistance, and postwar retribution and focusing on the Chetnik movement, this book analyses the politics of memory.

Since the overthrow of Slobodan Milošević in 2000, memory politics in Serbia has undergone drastic changes in the way in which the Second World War and its aftermath is understood and interpreted. The glorification and romanticisation of the Yugoslav Army in the Homeland, more commonly referred to as the Chetnik movement, has become the central theme of Serbia's memory politics during this period. The book traces their construction as a national antifascist movement equal to the communist-led Partisans and as victims of communism, showing the parallel justification and denial of their wartime activities of collaboration and mass atrocities. The multifaceted approach of this book combines a diachronic perspective that illuminates the continuities and ruptures of narratives, actors and practices, with in-depth analysis of contemporary Serbia, rooted in ethnographic fieldwork and exploring multiple levels of memory work and their interactions.

It will appeal to students and academics working on contemporary history of the region, memory studies, sociology, public history, transitional justice, human rights and Southeast and East European Studies.

Jelena Đureinović holds a PhD in Modern and Contemporary History from Justus Liebig University in Giessen, Germany, where she teaches in the Department of History. Her research deals with the history and politics of memory of the Second World War in Yugoslavia and the post-Yugoslav space with the focus on the process of reinterpretation of the Chetnik movement in Serbia. She was a visiting research fellow at the Moore Institute in Galway, the Centre for Southeast European Studies at the University of Graz and the Institute of Culture and Memory Studies in the Slovenian Academy of Sciences and Arts. She has published on Europeanisation and memory politics, memory laws, discourses of victimhood under communism and relations between memory cultures in Croatia and Serbia.

Southeast European Studies
Series Editor: Florian Bieber

The Balkans are a region of Europe widely associated over the past decades with violence and war. Beyond this violence, the region has experienced rapid change in recent times though, including democratisation, economic and social transformation. New scholarship is emerging which seeks to move away from the focus on violence alone to an understanding of the region in a broader context drawing on new empirical research.

The Southeast European Studies Series seeks to provide a forum for this new scholarship. Publishing cutting-edge, original research and contributing to a more profound understanding of Southeastern Europe while focusing on contemporary perspectives the series aims to explain the past and seeks to examine how it shapes the present. Focusing on original empirical research and innovative theoretical perspectives on the region, the series includes original monographs and edited collections. It is interdisciplinary in scope, publishing high-level research in political science, history, anthropology, sociology, law and economics and accessible to readers interested in Southeast Europe and beyond.

For more information about this series, please visit: www.routledge.com/Southeast-European-Studies/book-series/ASHSER1390

Post-Socialist Political Graffiti in the Balkans and Central Europe
Mitja Velikonja

The Politics of Memory of the Second World War in Contemporary Serbia
Collaboration, Resistance and Retribution
Jelena Đureinović

Social Mobilization Beyond Ethnicity
Civic Activism and Grassroots Movements in Bosnia and Herzegovina
Chiara Milan

The Politics of Memory of the Second World War in Contemporary Serbia
Collaboration, Resistance and Retribution

Jelena Đureinović

LONDON AND NEW YORK

First published 2020
by Routledge
2 Park Square, Milton Park, Abingdon, Oxon OX14 4RN

and by Routledge
52 Vanderbilt Avenue, New York, NY 10017

Routledge is an imprint of the Taylor & Francis Group, an informa business

© 2020 Jelena Đureinović

The right of Jelena Đureinović to be identified as author of this work has been asserted by her in accordance with sections 77 and 78 of the Copyright, Designs and Patents Act 1988.

All rights reserved. No part of this book may be reprinted or reproduced or utilised in any form or by any electronic, mechanical, or other means, now known or hereafter invented, including photocopying and recording, or in any information storage or retrieval system, without permission in writing from the publishers.

Trademark notice: Product or corporate names may be trademarks or registered trademarks, and are used only for identification and explanation without intent to infringe.

British Library Cataloguing-in-Publication Data
A catalogue record for this book is available from the British Library

Library of Congress Cataloging-in-Publication Data
A catalog record has been requested for this book

ISBN: 978-0-367-27804-5 (hbk)
ISBN: 978-0-429-29793-9 (ebk)

Typeset in Times New Roman
by Wearset Ltd, Boldon, Tyne and Wear

Giessen dissertation, Faculty of History and Cultural Studies (or FB04)

To Zoranka and Slavica, whose strength over the last decade has been my inspiration

Contents

Acknowledgements xi
List of abbreviations xiv

1 Introduction 1

Collaboration, resistance and retribution in Yugoslavia 2
Memory politics in post-Milošević Serbia: between
  anti-communism and ethnicisation 6
The context of the book: postsocialism 8
Memory and law: Serbia's pseudo-transitional justice 10
Approach 12
Outline 14

2 Exploring politics of memory 20

History of memory 20
Politicality of memory 22
State agency 23
Pluralities, struggles and layers 24
Post-Yugoslav Serbia as a case study 26

3 Yugoslav memory culture and its downfall 35

Yugoslav war memory 36
Main mnemonic agency 37
Reception of memory politics beyond compliance
  and rejection 41
Remembering the collaboration 42
When history outpoured 44

4 The Milošević era 52

Changes and continuities 53
The anti-communist opposition 55

x Contents

  Between the Chetnik revival and commemorations
   of the postwar retribution 57
  Ravna Gora gatherings 59

5 Memory politics in post-Milošević Serbia   65

  The war and its aftermath in the hegemonic narratives 66
  Purging Yugoslavia from the public 68
  National reconciliation: ending the civil war within the
   Serbian nation 70
  Imaginations of the Chetniks 72
  The Chetniks as victims of communism 79

6 Unearthing the past   86

  The Mihailović Commission 87
  The State Commission for Secret Graves 90
  The quest for the grave of Dragoljub Mihailović 100
  Informalities and failures of official fact-finding endeavours 104

7 Anti-communist memory politics from below   108

  Symbolic nature of state efforts 109
  Non-state actors 110
  Commemorative practices from below 115
  The symbolic power 122

8 History, memory and law   129

  Equalising the Chetniks and the Partisans 131
  Rehabilitation legislation 135
  Telling histories in the courtroom 137
  Judicial abolishment of the uprising 140
  Rehabilitation from below 142

9 Rehabilitation of Dragoljub Mihailović   147

  Agency behind the court case 148
  The Second World War in the courtroom 151
  Historians as expert witnesses 152
  Interventions against the process 155
  Discussing the 1946 trial 158
  Plaintiffs' claims summarised: the court decision 161

10 Conclusion   165

  Index   172

Acknowledgements

This book is the result of a five-year-long process. During my PhD, I was incredibly fortunate to work with Stefan Rohdewald and Vjeran Pavlaković as supervisors and I thank them for their interest, patience and believing in this project. I especially appreciate and remember the spontaneous meetings and informal chats about monuments Vjeran and I have had at countless different locations over the last few years. Paul Vickers often took a supervisor's role too, offering invaluable insights and comments that improved and nuanced my ideas. I could not be more grateful for all the time he invested in reading countless versions of my research proposal when I was at the very beginning and I enjoyed our ranting about the state of memory studies very much. Even though he is probably not aware of it, Kolja Lichy, who I worked with in the Department of East European History at the University of Giessen, contributed to shaping my approach by asking me a crucial question: "Have you thought about interviewing these people?"

This book would not be possible without the generous financial and institutional support I was fortunate to receive. Rosa Luxemburg Foundation in Germany funded my entire PhD journey and the Scholarship Foundation of the Republic of Austria supported my semester at the Centre for Southeast European Studies at the University of Graz. The fellowship at the Centre provided me with a much-needed productive atmosphere that marked the beginning of the writing process that results in this book. Here, I would especially like to mention Armina Galijaš and Dario Brentin, as well as my office mates Lura Pollozhani and Orhan Ceka. The travel grants of the International Graduate Centre for the Study of Culture financed the extensive fieldwork in Serbia and my time in the Research Area Cultural Memory Studies significantly shaped this book's approach. I would like to thank Jovana Mihajlović Trbovc and Tanja Petrović from the Institute of Culture and Memory Studies for inviting me to Ljubljana. Without the support of Bizzi and Bollo in the beginning, I would probably not be where I am today.

Staff at Matica srpska in Novi Sad, as well as the University and National Library in Belgrade, went out of their way to enable me to finish my research within the limited fieldwork time. Thanks to the openly accessible university libraries in Germany, I wrote most of the book in the draughty but stunning Social Sciences Library of the University of Hamburg.

xii *Acknowledgements*

Without my interviewees in Serbia and the time they dedicated to conversations with me, this project would not be complete. I thank them for the openness and readiness to discuss their work, activism and understanding of history with me, a person they did not know and did not have to meet. Some of them will not agree with my analysis of their work and our conversations and conclusions and I hope they will not take it personally. Although I do not focus on memory activism against historical revisionism in Serbia, Ida Boká, Dinko Gruhonjić, Nebojša Milikić, Milan Radanović, Miloš Perović and Nenad Porobić, among others, provided me with important insights into their struggle against the hegemony. In Belgrade and Novi Sad, special thanks go to fellow historians Milivoj Bešlin, Olga Manojlović Pintar, Srđan Milošević, again Milan Radanović, and Milan Ristović, conversations with whom, even if it was just once or twice, convinced me that this is a topic worth exploring and that my perspective is not wrong. I would also like to thank Florian Bieber and Robert Sorsby of Routledge for their interest in publishing this book.

On a more personal note, I am incredibly grateful for the many friendships that have been a part of this endeavour in studying memory politics. Katarina Kušić and Ljiljana Pantović made the research stays in Serbia not only more productive, but also a more entertaining experience. What also brought us together is that we examine different aspects of the same phenomenon and their work on non-military interventions and women's health significantly contributes to the general understanding of the postsocialist transformation in Serbia. Without the fieldwork in Belgrade, I would probably have never met Ruža, which would have been a tragedy. Nikola Baković, Magda Dolińska-Rydzek, Dora Komnenović and Stefan Trajković Filipović are among the people who had to listen to my endless ramblings about this project and I would like to explicitly thank them for that. A special shout out goes to Rodoljub Jovanović, who I have always considered a sort of partner in crime, even though we work on very different subjects.

Some people got to know me only in the final stages of my PhD and during the work on this book and I am especially grateful for their understanding, the strength to listen to all complaints and for making it easier, even if I was sometimes difficult. I would like to mention Ines, and Anna and Vera of my writing circle in Hamburg. The wonderful people in the archives of FC St. Pauli museum in Hamburg accepted me, made me feel welcome and put up with me for over a year. The rituals of archive Wednesdays and pre-match merchandise sale in front of the stadium were an inseparable part of the final writing stage of this manuscript. Forza!

I wish to thank my family without whose support I would never be able to embark on this journey. They were always there for me, at every stage of this project, even if they did not understand what I was doing, why it lasted so long and why was someone paying for it. It was quite literally a bumpy ride for us – when I visited Ravna Gora for the first time, it was my mother Zoranka who sat behind the wheel and battled the road holes. I dedicate this work to them, to Bojana, Zoranka, Slavica and Vinka, to *Đureinovići* and *Ubiparipovići*, as well as sorely missed Mišo and Rudi, who are not there to see this work finished but I know they would be proud. Last, but certainly not least, I owe a dept of gratitude to Ciarán, who was with me during the most stressful phases and, amazingly, never gave up his optimism.

Finally, I am very grateful to Damien Dempsey, whose work was the soundtrack to this book.

Abbreviations

AVNOJ	*Antifašističko veće narodnog oslobođenja Jugoslavije* (Antifascist Council of the People's Liberation of Yugoslavia)
BIA	*Bezbedonosno-informativna agencija* (Security Intelligence Agency)
DOS	*Demokratska opozicija Srbije* (Democratic Opposition of Serbia)
DS	*Demokratska stranka* (Democratic Party)
DSS	*Demokratska stranka Srbije* (Democratic Party of Serbia)
EU	European Union
NDH	*Nezavisna Država Hrvatska* (Independent State of Croatia)
NOB	*Narodnooslobodilačka borba* (People's Liberation Struggle)
NOR	*Narodnooslobodilački rat* (People's Liberation War)
OZNA	*Odeljenje za zaštitu naroda* (Department for People's Protection)
POKS	*Pokret obnove Kraljevine Srbije* (Movement for the Restoration of the Kingdom of Serbia)
RTS	*Radio televizija Srbije* (Radio Television of Serbia)
SDP	*Socijaldemokratska partija* (Social Democratic Party)
SPO	*Srpski pokret obnove* (Serbian Renewal Movement)
SPS	*Socijalistička partija Srbije* (Socialist Party of Serbia)
SRS	*Srpska radikalna stranka* (Serbian Radical Party)
SUBNOR	*Savez udruženja boraca Narodnooslobodilačkog rata* (Union of Associations of the Veterans of the People's Liberation War)
UDBa	*Uprava državne bezbednosti* (State Security Administration)

1 Introduction

In October 2014, Belgrade celebrated the 70th anniversary of its liberation in the Second World War by the Yugoslav Partisans and the Red Army. President of the Russian Federation Vladimir Putin visited Belgrade for the Victors' March, a spectacular military parade held on 16 October – four days before the actual Liberation Day – so that he could be present. Thousands of soldiers of the Serbian army took part in the parade, showcasing the air force, the river fleet, infantry units and equipment. National television broadcast the entire event live for several hours. The 65th anniversary generated numerous celebrations under the slogan "Belgrade Remembers", when President Boris Tadić invited another Russian president, Dmitri Medvedev, to celebrate together.

This newly invented tradition of pompous celebrations of liberation days and militarisation of commemorations became a regular practice in contemporary Serbia. The central celebratory event in 2017 took place at Batajnica Air Base near Belgrade, with the Russian minister of defence attending. Cities such as Novi Sad and Zrenjanin followed the example of the capital and have their own military parades. The Immortal Regiment procession is a massively growing practice that takes place across Serbia, appropriated from Russia and its diaspora.[1]

On all these occasions, Serbian officials emphasise the need to remember the struggle against fascism and "preserve that experience from the oblivion of history" (*RTS*, 2009). However, during the preparations for the 2014 military parade, Oliver Antić, the advisor of President Tomislav Nikolić, argued for including flags and symbols of the Chetnik movement. For him, "the Chetniks, the Partisans, the Russians and the Ukrainians were all on the same side" (Živanović and Jeremić, 2014), so there was no contradiction in the inclusion of everyone in the parade. The Chetniks, officially the Yugoslav Army in the Homeland, were the royalist armed forces led by Dragoljub Mihailović. They had Allied support until 1943 but they also engaged in collaboration. They were defeated by the Partisans and considered war criminals and collaborators in socialist Yugoslavia.

The initiative to reconcile the wartime enemies, the victorious Partisans and defeated Chetniks, failed and Chetnik flags did not flutter at the parade. A different commemorative tradition dedicated to the defeated forces of the Second World

War emerged in parallel to the Liberation Day festivities in the Serbian capital city. While Belgrade officially celebrates its liberation, a public holiday in Serbia, a significantly smaller community gathers in St. Sava Temple at a memorial religious service for victims of communist terror. Inviting "the citizens of Belgrade, Serbs, descendants of victims of bloody 20 October, the so-called liberation, and those who respect them", this community gathers to commemorate those who died in the reprisals by the Partisans at the end of the war (*Blic*, 2016). Here, the armed Second World War forces opposed to the Partisans transform into victims of communism, including the Chetniks, because their leader Mihailović was sentenced to death and executed in 1946.[2] State officials do not attend this annual memorial service and it does not attract significant media attention.

It might seem that Serbia embraces its Yugoslav and antifascist past and cherishes the Yugoslav memory culture based on the Partisans' victory, rather than seeking to disassociate itself from it. On the other hand, the religious counter-commemoration for victims of Yugoslav communism seems rather obscure. The truth is completely opposite from that, and far more complex. Belgrade did not always remember its liberation and the hegemonic discourses have stood closer to the memorial service for victims of communist terror than the celebration of the liberation by the communist-led Partisans.

After the overthrow of Slobodan Milošević in 2000, the Day of the Liberation was abolished, suiting the strong anti-communist political climate dominated by the understanding of the end of the Second World War as a communist occupation rather than liberation. The city renovated the Cemetery of Belgrade Liberators, previously neglected and left to decay, only for Medvedev's visit in 2009, when official commemorations reappeared. The return to commemorations culminated with the 2014 military parade and week-long celebrations, held a year after 20 October had returned to the official state calendar.

Strengthening political and economic relations with Russia is one of the main reasons for the sudden interest for the until recently completely abandoned holiday. The other reason is the prevailing of ethnicisation over anti-communism, the two lenses the official politics of memory in post-Milošević Serbia balances between. The revived commemorations decontextualise the liberation of Belgrade by nationalising, depoliticising and "de-Yugoslavising" it. The Yugoslav war memory transforms into the narrative of Serbian antifascism, and the common struggle of the Yugoslav and Soviet forces becomes an episode in the long history of Serbian and Russian alliance.[3]

Collaboration, resistance and retribution in Yugoslavia

The Second World War in Yugoslavia was both a civil war and a war of occupation and liberation, where different Yugoslav people "not only fought with or against the Axis forces, but they also fought against each other" (Sindbæk, 2012: 7). The Axis attack on 6 April 1941 marked the beginning of the war, after which the interwar Kingdom of Yugoslavia was dismembered into occupation zones. King Petar II and the government fled, and Yugoslavia capitulated.

Serbia was under the direct military command of the German occupation. The lack of staff made it necessary to rely on domestic collaboration (Prusin, 2017: 2), involving the Council of Commissars and the subsequent Government of National Salvation of Milan Nedić, both appointed in 1941. Their tasks were to maintain order and help with the exploitation of economic resources and the labour force. While they represented themselves as trying to save Serbia and Serbs from destruction, they directly participated in retaliatory measures against the very same population. Even though they justified collaboration as a necessary strategy without which the Serbian population would have suffered more devastation, there are no indications that "the collaborationist administration exerted any moderating influence on German policies in Serbia" (Prusin, 2017: 159).

The Chetniks were a nationalist resistance movement during the first months of the Second World War, before the conflict with the Partisans prevailed over anti-occupation struggles. The Chetniks amalgamated the incompatible: patriotism and betrayal; the strive for saving the nation and massive atrocities against it; heroism and utter cowardice; the strict military discipline and total ferociousness; and the alliance with the anti-Hitler coalition and the parallel collaboration with the Axis forces (Marjanović, 1979: 111). The Chetniks positioned themselves against the occupation, but they avoided fighting and attempted collaborating. This is one of the greatest contradictions of the Chetniks – considering the Axis were their enemies and at the same time collaborating with them against the Partisans. Mihailović wanted to collaborate "when he wanted, on his own terms and for his own purposes" (Ramet and Lazić, 2011: 30), and this is what made him a distrusted and unreliable partner. Anti-communism is crucial for understanding the decisions and activities of Mihailović and the Chetniks. They were ready to collaborate with anyone, to gain arms and ammunition for fighting the communists. Collaboration was never open because it was crucial to preserve the reputation of a resistance movement.

Although they represented the Yugoslav government, the Chetniks were predominantly Serbian. One of the important figures in the movement was Stevan Moljević, the author of the 1941 *Homogeneous Serbia* that advocated unification of all Serbs in an ethnically homogenous Greater Serbia. Although not officially a Chetnik program, the influence of Moljević and the idea of ethnic homogenisation is evident in their documents. As the end of the war approached, the Chetniks changed their orientation, replacing Moljević's homogeneous Serbia with a democratic and federal Yugoslavia.

The Chetnik units committed many crimes against civilians. Mihailović did not order them but was regularly informed and did nothing to stop it or punish the perpetrators, often not having control over commanders and units. The massacre in Vranić is one of the most prominent examples of Chetnik terror in Serbian villages. During one night in December 1943, the Avala Corps commanded by Spasoja Drenjanin Zeka killed 67 inhabitants of the village – women, the elderly and children, including a four-month-old baby. The targeted people were civilian family members of the Partisan fighters.[4] Crimes against Muslims and Croats were often a collective revenge for crimes against Serbs in the Independent State of Croatia.

While Mihailović was creating the first Chetnik strongholds in a small part of Serbia, the communist-led resistance movement had spread throughout Yugoslavia. The Partisans did not represent specific ethnic, nationalist or religious interests and were the only force rooted in the whole country that stood for territorial integrity and national equality of all the people of Yugoslavia (Manoschek, 1995: 123). Chetnik and Partisan units fought together in August and September 1941, even though their leaders could not agree upon a common strategy. The largest joint initiative of the two movements was the liberation of a large territory with around one million inhabitants – the Republic of Užice, ruled by power-sharing. The Chetnik attack on Partisan headquarters in the liberated territory started the conflict that would last until the end of the war. The German forces used the opportunity to go into a quick offensive, resulting in the fall of *Užička republika*.

A long-term partnership between the Partisans and Chetniks was impossible and irreconcilable differences in ideologies and future objectives made the conflict imminent. The possibility of compromise between the offensive tactics of the Partisans ("resistance at any costs") and the defensive tactics of the Chetniks ("protection of the Serbian nation from repression and waiting for the Allied offensive") was low (Sundhaussen, 2011: 258). Combining the strategy of armed resistance against the occupation and quislings with the revolution against the political and socioeconomic order of the Kingdom of Yugoslavia, the Partisans were a threat to the Chetniks as "defenders of the old regime" who wanted to keep the status quo (Tomasevich, 1975: 153).

The British support for the Chetniks included military and financial aid and propaganda that made the Chetniks very popular in Yugoslavia and abroad. In France, Mihailović became a legend as the leader of the resistance in the Balkans and Charles De Gaulle decorated him in 1943. This gave the Chetniks legitimacy, emphasised until today. The attitude of the western Allies shifted very gradually in 1942 and 1943, resulting in the abandonment of the Chetniks in favour of the Partisans at the 1943 Teheran Conference.

In November 1943, the Antifascist Council for People's Liberation of Yugoslavia (AVNOJ) established the foundations of socialist Yugoslavia, soon entering a compromise of provisional government with the government in exile that made the abandonment of the Chetniks official in 1944. King Petar II discharged Mihailović. In August 1944, Tito called all Chetniks and Home Guard units to join the Partisans, unless they had committed crimes against Yugoslav people. On 12 September, the king orders the Chetniks to put themselves under the command of Tito or they would be marked as traitors. Dragoljub Mihailović moved to Bosnia.

While the Chetniks from Herzegovina and Montenegro were retreating towards the West, Mihailović hoped to go back to Serbia and stayed in Bosnia, imagining an uprising against the communists. He still expected the western Allies to come, destroy communism and reinstate the monarchy, even after they recognised the new government of Yugoslavia (Lazić, 1997: 263). With the help of Nikola Kalabić, OZNA captured Mihailović in March 1946 and brought him to trial where he was sentenced to death.

The 1946 Belgrade Process was a public trial. Around 100 Yugoslav and foreign journalists followed the trial that was broadcast on Radio Belgrade. Transcripts were published the same year (*Izdajnik i ratni zločinac Draža Mihailović pred sudom: Stenografske beleške i dokumenta sa suđenja Dragoljubu-Draži Mihailoviću*, 1946). The public nature of the trial shows the intention "to stigmatise Mihailović as a national traitor in the widest possible sphere, inside as well as outside of Yugoslavia" (Sindbæk, 2012: 49). Mihailović's verdict "largely repeated the text of the indictment; again focus was on the collaboration and anti-Partisan activity of the Chetniks, while war crimes were included in a separate, minor chapter" (Sindbæk, 2012: 47). The trial reflected the official "view and representation that Draža Mihailović and the Chetniks were primarily traitors and enemies of the people, and secondarily war criminals and perpetrators of large massacres and ethnic cleansing" (Sindbæk, 2012: 47). Especially when observed from the perspective of today's legal standards, the Belgrade Process had many shortcomings, similarly to many postwar trials across Europe .

Towards the end of the Second World War and in the immediate postwar period, the Partisans and the Yugoslav authorities were settling accounts with their enemies. The retribution was predominantly aimed against those considered responsible for war crimes and collaboration. However, in the context of the aftermath of the civil war and the consolidation of state socialism, many people were punished unjustly or too harshly. Those who had collaborated with the occupation and fought against or persecuted the Partisans faced either extrajudicial executions or postwar trials. Confiscation of property usually followed judicial or extrajudicial measures, declaring a person a people's enemy (*narodni neprijatelj*).

The process of dealing with the collaboration in Yugoslavia reflected the wider European context at the time, where the first retributions, usually in the form of extrajudicial executions, happened around the war's end. The horrors of the occupation regimes led to brutalisation among the Partisans that manifested itself in revenge actions in which thousands of real or alleged enemies became victims (Zschächner, 2015: 60). The phase of retributive violence preceded institutionalisation through commissions and military tribunals, when the state apparatus took over the process of settling accounts with the wartime collaboration (Judt, 2005: 41–62). In Yugoslavia, the State Commission for the Investigation of Crimes of the Occupiers and Their Collaborators (*Državna komisija za utvrđivanje zločina okupatora i njihovih pomagača*, from now on: State Commission) formed as an investigative body. Numerous trials of different scales and publicity also took place.

Many Partisans' enemies fled the country at the end of the war, handing themselves over to the British soldiers in Austria and thinking they would avoid falling into Partisan hands. Milan Nedić attempted the same, but he was arrested and handed over to the Yugoslav authorities, who took him into custody and interrogated him before his trial in Belgrade. However, Nedić was never put on trial, because he died in February 1946, committing suicide by jumping through the window of the Belgrade prison, according to the official records.

Memory politics in post-Milošević Serbia: between anti-communism and ethnicisation

The Liberation Day commemorations illuminate the contradictory, inconsistent and often confusing nature of memory politics on the Second World War and state socialism in post-Milošević Serbia and sheds light on the interplay and oscillation between anti-communism and ethnicisation as prisms of interpretation of the Second World War and socialist Yugoslavia. This war interpretation has undergone drastic changes in Serbia, a post-Yugoslav, postsocialist and post-conflict society, especially since the overthrow of Milošević in 2000. The initial aim of dominant memory politics was separation from and delegitimisation of socialist Yugoslavia and its legacies. Othering and demonisation of Yugoslavia and everything it stood for, the memory culture included, would provide legitimacy to the new political elites that originated from the 1990s political opposition to Milošević. In the post-Milošević period, references to socialist Yugoslavia disappeared from the public sphere, the Axis collaborators have been recast as heroes and, directly linked to the former, a new social group of victims of communist terror emerged. This book explores this radical revision.

The Yugoslav interpretation of the Second World War centred on the People's Liberation War – the Partisans' struggle against the occupation and domestic collaborators – as the founding event and myth and main legitimacy source for Yugoslav state socialism. The post-Yugoslav memory politics inverted and replaced this war interpretation with the entirely opposite image of the Second World War and its actors. The narrative transformation particularly concerned the issues of collaboration, resistance and the immediate war aftermath, whose contemporary reading relates to executions, trials and repression, rather than the liberation from fascism. This parallels many of the processes in neighbouring Yugoslav successor states, such as Slovenia, Croatia and Bosnia and Herzegovina, where collaborators are often recast as national heroes to fit the new nation-building narratives.

The counter-narratives to the dominant war memory, that became hegemonic after 2000, had existed during socialist Yugoslavia. They were increasingly present in the public sphere in the 1980s, when the overall crisis provided a suitable context for them to surface and gain more popular support. During the rule of Milošević in the 1990s, the regime's ideology and memory politics were ambiguous, those in power not wanting to distance themselves from the so-called Partisan myth and representing themselves as the carriers of the People's Liberation War legacies. The political opposition commemorated the defeated political and military forces of the Second World War. The regime did not officially support them but also did not seriously restrict them, especially since these commemorations suited the broader atmosphere of nationalism in Serbia.

After the fall of Milošević, when his opposition came to power in the form of a very heterogeneous coalition, the previous counter-narratives could finally move to the official sphere. United in the anti-communist consensus and interpreting the Milošević era as the continuation of the communist regime, the political elites could frame themselves as liberators of Serbia from communism.

Anti-communism is the crucial driving force of positive recasting of the defeated movements of the Second World War and the abolishment of the Partisan myth. The roots for a very marginal and weak opposition to memory politics are in the anti-communist political consensus, characteristic for many postsocialist societies. This book historicises contemporary anti-communism in Serbia back to the late 1980s and early 1990s, demonstrating that it is the only common denominator of all opposition parties that emerged in this period, inscribed in their first party manifestos, in addition to the common goal of ousting Milošević. After these parties came to power, anti-communism became the state policy evident in many spheres beyond memory politics, including economy, education and healthcare. In a very similar manner, the sphere of anti-communist memory work from below involves politically diverse actors who cherish different images of the Chetniks and the past they commemorate, united in staunch anti-communism that brings them together as a community of memory.

Purging Yugoslavia and its memory culture from the public, the state authorities made reinterpretation of Second World War collaboration and resistance central to official memory politics, as it had been central to the Yugoslav regime. The transformation of the war memory intermixes with the narrative of the victimhood of the Serbian nation and state under communism. Through the emphasis on the postwar reprisals and "the crimes of the liberators", the Yugoslav Partisans, as politically unwanted ancestors, became perpetrators and their enemies turned into innocent victims of communism.

The metamorphosis of the defeated military and political movements of the Second World War into innocent victims of communism is accomplished by observing the entire Second World War through the lens of the postwar retribution, executions and trials. Such a selective view on the past implies justification, neglect and denial of wartime activities of the defeated, including collaboration, persecution of political opponents, war crimes and terror against civilians. This is how the Chetniks have become the national antifascist movement and victims of communism. The same lens applies to the Partisans, neglecting their struggle, the victory against fascism and emancipatory role for large segments of the population of Yugoslavia, reducing them to perpetrators hungry for political power. The gaze at socialist Yugoslavia through the prism of its most repressive period remoulds the decades of state socialism into an exclusively negative historical experience.

The Partisans can only appear as a positive historical reference for the Serbian nation-state and its elites through ethnicisation of the Second World War and framing their struggle and victory as Serbian. The ethnonationalist lens is equally as narrow as the anti-communist one and deprives the People's Liberation Movement from its communist ideology and Yugoslav and multi-ethnic dimension. Ethnicisation goes hand in hand with the national reconciliation narrative that equalises the Chetniks and the Partisans as two Serbian antifascist movements.

Even though it seemingly does not fit the anti-communist consensus, ethnicisation often intertwines with memory politics driven by anti-communism. This trend started with the first official Liberation Day celebrations at the end of the first post-Milošević decade, when the Democratic Party (*Demokratska stranka*, DS) was still in power and Boris Tadić was the president of Serbia. The revival of state-sponsored antifascist commemorations equally relates to the European Union accession and always closer relations with the Russian Federation. As the Serbian Progressive Party (*Srpska napredna stranka*, SNS) rose to the political hegemony from 2012 on, with Milošević's Socialist Party of Serbia (*Socijalistička partija Srbije*, SPS) as the crucial partner, nationalisation of the Second World War achieved mastery over anti-communism. While embracing the Partisans as the proof that the Serbian nation has always been on the right side of history, the political elites do not separate themselves from the political and legal rehabilitation of the Chetniks, a hallmark of memory politics in the post-Milošević period.

The context of the book: postsocialism

The fall of communism in Central and Eastern Europe generated a push for addressing the communist past through state-sanctioned memory politics. At first, this process involved drawing a "thick line" of forgetting under the period, as was the case in the first postsocialist years in Poland, and Central and East European countries did not commemorate the victory over communism. It was only in the late 1990s that the doubt surfaced that communism had not really been defeated, leading to initiatives of engaging with the communist past in order not to confirm the mastery "over a now demonised and banished political system, but to confront and weaken its continuing hold on the present" (Mark, 2010: xiii–iv).

Memory practices adopted by postsocialist political elites after 1989 were a continental novelty (Mark, 2010: xiii). Without a doubt, there are similarities in the majority of the postsocialist states, where memory of communism is intertwined with memory of the Second World War, as it had provided legitimacy to the communist rule for decades. After the fall of communism, "a new cleansing of the past took place in which the formerly dominant class approaches to 'resistance' and 'collaboration' were replaced with a focus that often reflected revisionist tendencies" (Karge, 2010: 138). In most postsocialist countries, instead of the expected and idealised democratic pluralisation of memory that would follow the fall of communism, powerful institutions have been employed to forge a unitary interpretation of the past from above (Mark, 2010: 59). In this sense, there are commonplaces across the vast geographical region defined by the historical experience of decades of communism.

Taking the similarities in the historical experience of the Second World War, state socialism and still ongoing transition into consideration, the post-Yugoslav context involves numerous specificities. The fall of Yugoslav communism involved state disintegration and armed conflicts that continued even into the twenty-first century (Ostojić, 2014: 21). The transition in the Yugoslav case was not only a postsocialist transition into liberal democracy, "but to nationalism, war and violence" (Subotić, 2015: 188). The fall of communism was not something to be celebrated as it led to "protracted civil war, grim levels of violence and ultimately genocide" (Subotić, 2015: 189). In Serbia, the authoritarian regime of Slobodan Milošević persisted until 2000.

Yugoslav state socialism was unique in many ways. It enjoyed massive popular support that gained momentum during the Second World War and stretched across the ethnic boundaries. For large segments of society, it was an emancipating experience. The external factors such as the Red Army were not the main actors decisive for the Yugoslav revolution that transpired in the wartime circumstances. The split with the Soviet Union in 1948 and the position of a co-founder and unofficial leader of the Non-Aligned Movement informed the Yugoslav foreign policy of balancing between the blocs, or, more precisely, being both in the East and in the West (Vučetić, 2018: ix). After the initial period of repression during the regime consolidation, Yugoslavia underwent the process of liberalisation and decentralisation, as well as numerous phases of crisis. It remained a one-party state until the early 1990s.

The phenomenon of "transnationalism in reverse" (Kirn, 2016) in memory culture is an aspect of the transformation process from a large-scale multinational state to smaller and predominantly nation-states. The transnational narrative of the Second World War as the common struggle of all Yugoslav people against the occupation went through the process of national fragmentation of memory into divided and divisive nation-centred memory cultures. During the wars of the 1990s and continuing in the decades afterwards, the history of the common Yugoslav state became a burden, an outfit that became uncomfortable, too tight and not fashionable anymore, and it was restricting the movement of the growing ethnocentric and nationalist tendencies. Thus, it had to be thrown away (Ristović, 2013: 139).

My intention is not to argue that there is absolute exceptionalism that separates Yugoslavia and the post-Yugoslav context from other countries with the communist past. I rather emphasise specificities to underline the need to fragment large research paradigms with locatedness in mind (Radstone, 2011). In other words, understanding travelling of memory and its carriers, modes, contents, forms and practices as transnationally constituted does not necessarily imply a homogenising approach that moves away from the locatedness of history and memory cultures (Erll, 2011: 11). The focus on national frameworks and specificities such as the Serbian case can help fragment often homogenising global, European, East European and postsocialist perspectives and foster more in-depth comparisons with implications for understanding the phenomenon of memory politics in general.

While emphasising the importance of locatedness of research, I use the category of "postsocialist" when talking about Serbia, aware of the question how long it will be tenable to speak, write and think of a category called "postsocialism" (Mandel and Humphrey, 2002: 3). The reason why Serbia is postsocialist is because postsocialism remains "a pertinent concept in a context in which transition lives on as an unfinished business" (Mikuš, 2018: 7–8). Understanding that it will not be the case forever, I consider "postsocialist" to be relevant because the legacies of socialism still significantly affect the societies once ruled by it (Humphrey, Hann and Verdery, 2003). For similar reasons, I refer to Serbia as post-Milošević because the legacies of this era are still very tangible in society.

Memory and law: Serbia's pseudo-transitional justice

The book dedicates special attention to law because of the important role law plays in memory politics as its instrument and mirror. The legal sphere also acts as a contact zone where different levels of memory work meet. Naturally, legal means constantly regulate the past and the main frameworks of official memory politics are always rooted in law, including memorial days, monuments, museums and school curricula. Mediated trials and specific laws regulating the memory of the Second World War and its atrocities have existed across Europe since the postwar period. The decommunisation paradigm in Central and Eastern Europe and transitional justice processes of recent decades have contributed to the emergence of legislation with a specific purpose of addressing the past, usually referred to as memory laws.

The question is not whether law shapes memory because it always does so. The question is rather how law performs that task (Heinze, 2016: 8). Law "validates those past events that we approve of and invalidates those of which we don't: by honouring past promises, punishing past wrongdoings, by rehabilitating victims or by offering compensation for losses" (Markovits, 2001: 514). Law became crucial in processes of constructing the past as a series of wrongs that need to be "rectified", on the one hand, and a tool with which these injustices can allegedly be repaired, on the other (Loytomaki, 2014: 9). The relationship between law and memory is reciprocal (Savelsberg and King, 2007).

The concept of memory laws is a rather recent phenomenon in scholarship, although such laws have been increasingly passed since the postwar period (Koposov, 2017). Memory laws represent a special category of law created specifically for the purpose of regulating historical interpretation, whether by recognition, evaluation, prescription, sanction or prohibition. In this sphere, we can distinguish between "a narrow category of specifically dedicated memory laws and, on the other hand, the inherent and far broader phenomenon of laws affecting public memory" (Heinze, 2016: 31).

A large part of this book is dedicated to the question of how law performs the task of shaping memory. I incorporate legislation that could be considered memory laws into analysis, such as the law on official holidays, with the focus on legislation that formally does not belong to this category. The laws the book concentrates on are rather examples of law affecting public memory that reflect the transnational paradigm of "judicialization of history" (Evans, 2002: 326). These are laws on veteran rights and rehabilitation of victims of political repression and violence (*Službeni glasnik Republike Srbije*, 2004; *Službeni glasnik Republike Srbije*, 2006; *Službeni glasnik Republike Srbije*, 2011).

Veteran and Rehabilitation Laws have very specific practical purposes, clearly defined in the legal text – regulation of veteran rights and benefits and judicial rehabilitation procedure. Such legislation exists in most legal systems and does not have to relate to a specific historical experience, including state socialism. Formally, the function of these laws is not a regulation of historical interpretation. However, they are a vital part of memory politics in Serbia and promote certain historical narratives that reflect the hegemonic discourses. The Veteran Law recognised the Chetniks as an antifascist movement and Rehabilitation Laws generate one-sided court processes that reach the public and demonise the Partisans and Yugoslav regime as criminal. They both incorporate the defeated political and military actors of the Second World War and serve their positive recasting.

When it comes to postsocialism, many studies of memory laws and the relationship of law and memory analyse the level of success or failure of these measures but do not challenge the revisionist narratives behind it, often starting with the assumption that state socialism was generally and exclusively a negative historical experience that should be dealt with through legal measures. The most detailed legal analyses that assess the political instrumentalisation of the communist past critically focus on contemporary Russia, seeing its legal efforts as faux or pseudo-transitional justice (Andrieu, 2011).[5]

Similarly to Russia, the efforts of the Serbian political elites aiming at overcoming the communist past are pseudo-transitional justice, "a tool for helping the government uphold a historical narrative favourable to it" (Andrieu, 2011: 213). They do not really aim at victims' rights but should foster legitimacy to new political elites. This is certainly the case with the decommunisation in Central and Eastern Europe in general, where memory laws and very specific postsocialist models of institutes of national remembrance and museums of communist victimisation equally serve the political and legitimacy purposes. Post-communist states strategically use political memory to resolve their contemporary ontological insecurities (Subotić, 2018: 297). Nation's victimhood under communism is the central narrative in post-communist societies and communism symbolises the evil, while the Holocaust memory, central to Europe, threatens and destabilises their state identities (Subotić, 2018: 300).

Serbia's laws, fact-finding commissions and other efforts are a perfect example of pseudo-transitional justice focused on communism, while neglecting the 1990s wars as the more urgent past that should be dealt with. Institutional and legal mechanisms that should accommodate and compensate the victims of communism are instruments of criminalisation and delegitimisation of the People's Liberation War and Yugoslav state socialism. Revising the history of the Second World War and socialist Yugoslavia, they simultaneously construct new victims and new perpetrators. The victimhood and all faux transitional justice efforts revolve around the defeated side of the Second World War. In the hierarchy of those considered victims of communism, who have the right to have their human rights violations of the distant past acknowledged, the Chetniks take the highest and pivotal position. Their commander, Dragoljub Mihailović, is the central figure of the misuse of transitional justice mechanisms and human rights discourses, with a state-funded commission dedicated only to investigation of his death and burial site.

12 *Introduction*

All-encompassing mnemonic hegemony in Serbia did not result in concrete accommodation of victims of communism, their straightforward recognition through an official memorial day, monument or state-sponsored commemorative practices. The existence of some mechanisms that resemble transitional justice and human rights discourses that political elites promote should not be mistaken for the actual concern for victims of communism and their descendants. In dominant memory politics, they only represent rhetorical and political symbols and instruments of historical revisionism.

The neglect by the state is the reason why non-state actors organise themselves to commemorate the past, how they want it to be remembered, and honour those they consider victims of communism. The anti-communist memory work from below bears resemblance to the institutional level in continuities with the 1980s, centrality of the Chetniks and anti-communism as the unifying factor. Frustrated with state efforts, anti-communist memory communities are in a constant struggle for recognition, but they also inform and shape the official memory politics through participation in its mechanisms, contributing to its informal and frivolous nature.

Approach

This book seeks to illuminate how politics of memory works. It offers an in-depth and historically contextualised analysis of memory politics in Serbia, as it relates to Second World War collaboration and resistance and the immediate postwar period, taking the Chetnik movement as the main case study. Such analysis requires "a chameleon-like ability to adopt a multitude of perspectives" (Berger and Niven, 2014: 12). The twofold approach, fruitful for its potential application to other contexts, combines a diachronic perspective that traces developments, transformations, continuities and ruptures of narratives, practices and actors, with an in-depth investigation of the contemporary context, actor-centred and encompassing multiple levels of memory work. Besides the diverse types of primary sources this book relies upon, I made use of the contemporary nature of the research subject and my general leanings towards ethnography and conducted interviews with relevant actors while attending and observing commemorations and other events.

Politics of memory is not limited to state agency and actors in power positions, it is a multi-sided struggle for hegemony that does not imply a dichotomy of imposed narratives from above and opposition from below. Mnemonic agency from below can replicate the hegemonic discourses of political elites, while seeking recognition and memory activism does not have to be liberal or leftist. The concept of multi-level memory work coined by Brian Conway is particularly useful for understanding how memory politics really works (Conway, 2010).

Hegemonic discourses in Serbia resonate at all levels of memory work from below, which informs back the mechanisms of state-sponsored memory politics. In this book, I focus on the understanding of anti-communist memory communities, seeking to grasp how dominant memory resonates in society and going beyond the romantic dichotomy of state agency and opposition. While the opposition to historical revisionism in Serbia exists in different forms, usually coming from the liberal and leftist positions, it is marginal in public and in the sphere of official memory politics characterised by the anti-communist consensus. For instance, there are no recent films that critically engage with the dominant memory and treat the Chetniks in a counter-hegemonic way, as opposed to numerous television programs about communist terror and Chetniks as ideal ancestors. The same applies to publicly funded museums and other public representations of history this book scrutinises.

While the book has a very clear focus on a national context, it brings the Serbian case into communication with the wider setting. One of the most obvious ones is the post-Yugoslav space. As Dora Komnenović underlines,

> as much as former Yugoslav countries would like to distance themselves from one another, gloating over 'misdeeds' of a neighbour is often done in an all-Yugoslav frame of reference with the purpose of cleaning up one's reputation and possibly relativising one's own faults.
>
> (2018: 195)

Gloating is a crucial segment of the mnemonic dynamics between Serbia and Croatia, especially concerning the memory of the Second World War. The accusations of the other side being more perpetrating in the past and revisionist in the present build the core of the debates between the two countries. Interestingly, they also share countless similarities, including anti-communism and national reconciliation of the victorious and the defeated of the Second World War.

Chetniks as ideal antecedents

What makes the Chetniks a fascinating case and easy object of revision is their ambivalent nature. They were a resistance movement supported by the Allies, while collaborating with the occupation and quisling forces. As the army officially representing the Yugoslav government in exile, they were extremely disorganised and engaged in mass atrocities against civilians. Their specific and equivocal characteristics enable their various reinterpretations.

Serbia's memory politics is characterised by the Chetnik centrality. The book traces the construction of the Chetniks as the national antifascist movement equal to the Partisans and as victims of communism, examining how the construction of the Chetnik myth relates to the denial and justification of their collaboration and crimes and the parallel deconstruction of the Partisan myth. With the narratives of national reconciliation and victimisation at the forefront, the Chetniks became the ideal historical reference precisely because they can be constructed as both an antifascist movement and innocent victims of communism. According to this view, the Chetniks fought against both totalitarianisms and ended up as victims of the communist one. This narrative demands leaving out the issues of extreme nationalism, collaboration and often ethnically motivated crimes against civilians, together with all other facts that do not suit their positive remoulding.

Antifascism, anti-communism and victimhood form the core of the positive image of the Chetniks that dominates today. This image is hegemonic but not homogeneous. It is as diverse as the actors who promote it – from the Chetniks as the Yugoslav democrats admiring Western democracies to celebrations of their plans for ethnically homogeneous Serbia, deeply intertwined with the 1990s wars. This argument of the book represents a challenge to the stereotype dominant outside of Serbia and within the opposition to state-sponsored historical revisionism, that sees the political rehabilitation of the Chetniks as a one-dimensional phenomenon.

Outline

Chapter 2 outlines the theoretical and methodological contribution of this book that combines a diachronic and empirical approach while understanding memory politics as a multi-layered phenomenon. It outlines the main concepts, dedicating attention to the notion of memory work, its multi-level nature and actors, levels and hierarchies. The second section of the chapter discusses the implications of doing research on an ongoing phenomenon and how the contemporary nature of the subject influences the approach. It elaborates on studying memory politics in post-Yugoslav Serbia, addressing the state of the sources the book relies upon, the different paths towards accessing them and the problems that arise in the process. The closing section reflects on interviews and fieldwork conducted in Serbia.

Chapter 3 explores the history of memory on the Second World War in socialist Yugoslavia from the postwar period to the end of the 1980s, focusing on the Yugoslav memory politics and culture and their crumbling after the death of Tito. The chapter problematises the straightforward readings of the Yugoslav memory culture and demonstrates that people interacted diversely with the official frameworks of memory, although the Partisan myth dominated the memory culture. The public representations of the Partisans, the Chetniks and the historical experience of socialist Yugoslavia underwent transformations during the political, economic and legitimacy crisis of the 1980s. The narratives that surfaced in this period became the backbone of historical revisionism about the Second World War and Yugoslavia in postsocialist Serbia.

The following chapter investigates the period of Slobodan Milošević's rule in "rump" Yugoslavia in the 1990s. The Chetnik commemorations emerged in the early 1990s within the political opposition to Milošević that had anti-communism as a common denominator. At the same time, the regime employed ambiguous memory politics, promoting itself as representing the continuity with the Yugoslav Partisans in combination with nationalism, while tolerating the commemorations of the defeated Second World War forces. The mixture of ethnonationalism and the Partisan myth narrated the war through the dichotomy of Serbian victimhood and heroism. Within the opposition, common commemorative practices such as the Ravna Gora gathering and religious memorial services emerged in this period. Different understandings of the characteristics of the Chetniks and positions towards the wars of the Yugoslav succession caused the division within this memory milieu that continued after 2000, when Milošević's opposition came to power.

Chapter 5 explores the dominant politics of memory in post-Milošević Serbia. The drastic revision of the Second World War that became the central feature of memory politics is rooted in the anti-communist consensus of the state actors. This chapter examines the different phases of the institutional level work, including the initial attempts to purge Yugoslavia from the public sphere and the national reconciliation narrative that not only serves the purpose of rehabilitating the Chetniks, but also portrays the Partisans as having started the civil war within the Serbian nation. The chapter focuses predominantly on the memory politics from above and analyses parliamentary debates, commemorations, television series and museum representations of history, while focusing on the representations of the Chetniks. The chapter illuminates the Chetniks as taking the highest position in the hierarchy of victims of communism, with Dragoljub Mihailović as the symbol of the victimhood of the entire movement.

Chapter 6 moves on to discuss the work and results of the two fact-finding commissions dealing with the postwar period, illuminating the informal nature of the institutional level of memory work and underlining that the Chetniks and Dragoljub Mihailović constitute the central theme of official memory politics in contemporary Serbia. The State Commission for Secret Graves and the commission dedicated to investigating circumstances of Mihailović's execution are the main subjects of analysis. Established by the government, the commissions did not resemble a state-funded investigative body, but they often seemed like informal endeavours. Because of this oscillation between an official governmental body and a bottom-up initiative, the sensationalist media reports and the failure to fulfil their goals, the two commissions represent a fascinating case study for the wider dynamics of state-sanctioned memory politics in Serbia.

By looking at the sphere of anti-communist memory work from below, Chapter 7 focuses on the actors operating at the small-group and social level of memory work while seeking recognition by the state. Although this sphere gathers non-state actors who are usually not in positions of power, certain holders of symbolic power and authority provide their support to commemorative practices, namely the Karađorđević family and the Serbian Orthodox Church. The chapter scrutinises the agency of this level of memory work and how these actors perceive socialist Yugoslavia, the Milošević era and the political changes that transpired after his overthrow. Furthermore, I investigate the practices, including commemorations and memorials for victims of communism and the defeated forces of the Second World War, paying close attention to the Chetnik commemorations. Finally, I trace the struggles for recognition that constitute the essential dimension of the anti-communist memory work from below.

16 *Introduction*

Law has been an especially relevant dimension of official memory politics in Serbia since the overthrow of Slobodan Milošević. Chapter 8 examines law as an instrument and a mirror of politics of memory, tackling the cases of the revised veteran legislation that equalised the Chetniks and the Partisans and the mechanism of judicial rehabilitation of victims of communism. Acknowledging that the Yugoslav authorities also unjustly persecuted people in the postwar period, the chapter is primarily concerned with the military and political actors of the Second World War, including Dragoljub Mihailović and Milan Nedić. Based on an analysis of legislation and available court transcripts and observations of court proceedings in numerous processes of rehabilitation in Serbia, this chapter understands the legislation and practice of rehabilitation as a legal confirmation of the historical narratives already dominant in other spheres. Rehabilitation processes also serve as contact zones where different levels of memory work intertwine.

The final chapter examines the process of judicial rehabilitation of Chetnik commander Dragoljub Mihailović that ended with a positive court decision in 2015. The significance of this case is that it represents the formal epilogue and a legal confirmation of the decades-long process of positive reinterpretation of the Chetnik movement, intertwined with the narratives of their victimhood, that has been promoted at multiple levels of memory work in Serbia. While the court case and its outcome revolved around the hegemonic discourses about the Second World War and state socialism, it was the agency from below that initiated the process, namely the plaintiffs from the anti-communist memory community. The apparent personal background and motivation for judicial rehabilitation is not only intertwined with the political, but the political dimension prevails. The chapter addresses numerous issues of the rehabilitation process, including the problem of selective interpretation of history and the participation of historians as expert witnesses.

Notes

1 In Republika Srpska and Montenegro as well. Immortal Regiment procession is a practice where people carry photos of their relatives who fought in the Second World War. It emerged in Siberian city Tomsk in 2012 and quickly spread across the Russian diaspora worldwide.
2 Many Chetniks died in combat against the Partisans at the end of the war or faced executions, but a large part was given amnesty in 1944 or managed to flee Yugoslavia, similarly to many other Partisans' adversaries such as the Ustasha in Croatia or the Slovene Home Guard.
3 The official celebrations of 9 May promote the same discourse. The First World War appears as a reference at Second World War commemorations as politically more suitable. When President Boris Tadić came to the stage during the 2009 commemoration, *Marš na Drinu* (March to the Drina) played in the background. The same song was the grand finale of the 2014 Victors' March parade. Composed by Stanislav Binički during the First World War, the song celebrates the bravery of the Serbian people and the victory in the 1914 Battle of Cer. The lyrics were added in 1964 and the song was very popular during socialist Yugoslavia.

4 The High Court in Belgrade rehabilitated Drenjanin in 2015 as deprived of life and property for political reasons. On the crimes of collaborationist forces, including the Chetniks, see: Radanović, M. (2015) *Kazna i zločin. Snage kolaboracije u Srbiji*. Belgrade: Rosa Luxemburg Stiftung.
5 As opposed to the legal studies and transitional justice paradigm, many authors studying postsocialist memory politics do critically approach the communist victimisation narrative and its political instrumentalisation. See, for instance: Clarke, D. (2014) "Communism and Memory Politics in the European Union", *Central Europe* 12(1): 99–114. Ghodsee, K. (2014) "A Tale of 'Two Totalitarianisms': The Crisis of Capitalism and the Historical Memory of Communism", *History of the Present* 4(2): 115–142. Neumayer, L. (2018) *The Criminalisation of Communism in the European Political Space after the Cold War*. Routledge. DOI: 10.4324/9781351141765.

Bibliography

Andrieu, K. (2011) "An Unfinished Business: Transitional Justice and Democratization in Post-Soviet Russia", *International Journal of Transitional Justice* 5(2): 198–220. DOI: 10.1093/ijtj/ijr011.

Berger, S. and Niven, B. (2014) "Introduction". In Berger, S. and Niven, B. (eds), *Writing the History of Memory*. London: Bloomsbury Academic, pp. 1–24.

Blic (2016) "Parastos žrtvama komunizma sutra u Hramu Svetog Save". Available at: www.blic.rs/vesti/drustvo/parastos-zrtvama-komunizma-sutra-u-hramu-svetog-save/2wsq42h (accessed 17 March 2018).

Conway, B. (2010) *Commemoration and Bloody Sunday: Pathways of Memory*. Palgrave Macmillan Memory Studies. Basingstoke: Palgrave Macmillan UK.

Erll, A. (2011) "Travelling Memory", *Parallax* 17(4): 4–18. DOI: org/10.1080/13534645.2011.605570.

Evans, R. J. (2002) "History, Memory, and the Law: The Historian as Expert Witness", *History and Theory* 41(3): 326–345. DOI: org/10.1111/1468-2303.00209.

Heinze, E. (2016) *Beyond "Memory Laws": Towards a General Theory of Law and Historical Discourse*. ID 2797238, SSRN Scholarly Paper, 17 June. Rochester, NY: Social Science Research Network. Available at: https://papers.ssrn.com/abstract=2797238 (accessed 28 March 2017).

Humphrey, C., Hann, C. M. and Verdery, K. (2003) "Introduction: Postsocialism as a Topic of Anthropological Investigation". In Hann, C. M. (ed.), *Postsocialism: Ideals, Ideologies and Practices in Eurasia*. London: Routledge, pp. 1–29.

Izdajnik i ratni zločinac Draža Mihailović pred sudom: Stenografske beleške i dokumenta sa suđenja Dragoljubu-Draži Mihailoviću (1946) Belgrade: Savez udruženja novinara Federativne Narodne Republike Jugoslavije.

Judt, T. (2005) *Postwar: A History of Europe Since 1945*. New York: The Penguin Press.

Karge, H. (2010) "Practices and Politics of the Second World War Remembrance: (Trans-)national Perspectives from Eastern and Southeastern Europe". In Pakier, M. and Stråth, B. (eds), *A European Memory? Contested Histories and Politics of Remembrance*. New York: Berghahn Books, pp. 137–147.

Kirn, G. (2016) "Transnationalism in Reverse: From Yugoslav to Post-Yugoslav Memorial Sites". In de Cesari, C. and Rigney, A. (eds), *Transnational Memory. Circulation, Articulation, Scales*. Media and Cultural Memory 19. Berlin: De Gruyter, pp. 313–338.

Komnenović, D. (2018) "The 'Cleansing' of Croatian Libraries in the 1990s and Beyond or How (Not) to Discard the Yugoslav Past". In Bevernage, B. and Wouters, N. (eds), *State-Sponsored History Since 1945*. London: Palgrave Macmillan, pp. 191–207.

Koposov, N. (2017) *Memory Laws, Memory Wars: The Politics of the Past in Europe and Russia*. Cambridge: Cambridge University Press.
Lazić, M. (1997) *Ravnogorski Pokret*. Belgrade: Institut za noviju istoriju Srbije.
Loytomaki, S. (2014) *Law and the Politics of Memory: Confronting the Past*. Routledge.
Mandel, R. and Humphrey, C. (2002) *Markets and Moralities: Ethnographies of Postsocialism*. Oxford, New York: Berg.
Manoschek, W. (1995) *"Serbien ist judenfrei": Militärische Besatzungspolitik und Judenvernichtung in Serbien 1941/42*. Berlin: De Gruyter.
Marjanović, J. (1979) *Draža Mihailović između Britanaca i Nemaca*. Zagreb, Beograd: Globus, Narodna knjiga.
Mark, J. (2010) *The Unfinished Revolution: Making Sense of the Communist Past in Central-Eastern Europe*. London: Yale University Press.
Markovits, I. (2001) "Selective Memory: How the Law Affects What We Remember and Forget about the Past: The Case of East Germany", *Law & Society Review* 35(3): 513–563. DOI: 10.2307/3185395.
Mikuš, M. (2018) *Frontiers of Civil Society: Government and Hegemony in Serbia*. New York, Oxford: Berghahn Books.
Ostojić, M. (2014) *Between Justice and Stability: The Politics of War Crimes Prosecutions in Post-Miloševic Serbia*. Surrey: Ashgate.
Prusin, A. (2017) *Serbia under the Swastika: A World War II Occupation*. Urbana: University of Illinois Press.
Radstone, S. (2011) "What Place Is This? Transcultural Memory and the Locations of Memory Studies", *Parallax* 17(4): 109–123. DOI: org/10.1080/13534645.2011.605585.
Ramet, S. P. and Lazić, S. (2011) "The Collaborationist Regime of Milan Nedić". In Ramet, S. P. and Listhaug, O. (eds), *Serbia and the Serbs in World War Two*. Basingstoke: Palgrave Macmillan UK, pp. 17–44.
Ristović, M. (2013) "Kome pripada istorija Jugoslavije?" *Godišnjak za društvenu istoriju* XX(1): 133–143.
RTS (2009) "Dan oslobođenja Beograda". Available at: www.rts.rs/page/stories/sr/story/125/drustvo/239281/dan-oslobodjenja-beograda.html (accessed 3 March 2018).
Savelsberg, J. J. and King, R. D. (2007) "Law and Collective Memory", *Annual Review of Law and Social Science* 3(1): 189–211. DOI: org/10.1146/annurev.lawsocsci.3.081806.112757.
Sindbæk, T. (2012) *Usable History? Representations of Yugoslavia's Difficult Past from 1945 to 2002*. Aarhus: Aarhus University Press.
Službeni glasnik Republike Srbije (2004) Zakon o pravima boraca, vojnih invalida i njihovih porodica 2004.
Službeni glasnik Republike Srbije (2006) Zakon o rehabilitaciji.
Službeni glasnik Republike Srbije (2011) Zakon o rehabilitaciji.
Subotić, J. (2015) "The Mythologizing of Communist Violence". In Stan, L. and Nedelsky, N. (eds), *Post-Communist Transitional Justice: Lessons from Twenty-Five Years of Experience*. Cambridge: Cambridge University Press, pp. 188–210.
Subotić, J. (2018) "Political Memory, Ontological Security, and Holocaust Remembrance in Post-communist Europe", *European Security* 27(3): 296–313. DOI: org/10.1080/09662839.2018.1497980.
Sundhaussen, H. (2011) "Besetzte jugoslawische Gebiete Kroatien, Serbien, Montenegro und Bosnien-Herzegowina". In Ueberschär, G. R. (ed.), *Handbuch Zum Widerstand Gegen Nationalsozialismus Und Faschismus in Europa 1933/39 Bis 1945*. Berlin: De Gruyter, pp. 255–269.
Tomasevich, J. (1975) *The Chetniks*. Stanford: Stanford University Press.

Vučetić, R. (2018) *Coca-Cola Socialism: Americanization of Yugoslav Culture in the Sixties*. Budapest: CEU Press.
Živanović, K. and Jeremić, V. (2014) "Antić: Pravedno da obeležja četnika budu na paradi". Available at: www.danas.rs/drustvo/antic-pravedno-da-obelezja-cetnika-budu-na-paradi/ (accessed 17 March 2018).
Zschächner, R. (2015) *Kein Ende der Geschichte: Diskurse, Gedenken und Formen der Erinnerung an den Zweiten Weltkrieg in Serbien*. Berlin: Weißensee Verlag.

2 Exploring politics of memory

This book takes two parallel approaches to memory politics on the Second World War, state socialism in the post-Yugoslav space and a reinterpretation of the defeated side of the Second World War in Serbia. First, as a historian, I am interested in the diachronic dimension of war memory and its longue durée developments and transformations of historical narratives since the postwar period. In this sense, this book represents the history of the parallel processes of unmaking and making of the Partisan and Chetnik myths. The book complements the diachronic perspective with a focus on the dynamics of memory politics in contemporary Serbia. A multi-layered and empirically grounded approach illuminates that there is more to memory politics than a top-down level encompassing state agency, institutions and groups in power positions and that mnemonic hegemony does not necessarily imply homogeneity.

History of memory

There are no transhistorically valid answers to the question of how groups and societies remember their past, but acts of remembering have their own specific history (Mark, 2010: xiii). Writing about the history of memory can illuminate the historical variability of specific constellations of collective remembering, as well as of the very concepts of history and memory (Erll, 2011: 45). Studying memory as a historical phenomenon is what Stefan Berger and Bill Niven (2014a) simply term the history of memory, as a branch of historiography. For Peter Burke, this is the social history of remembering (1997: 46). With the understanding that social memory is selective, as individual memory, historians of memory need to identify the principles of selection and examine how they differ from place to place or from group to group and how they change over time. As "memories are malleable", historians need to understand how they are shaped and by whom and examine the limits to this malleability (Burke, 1997: 46).

In his study of the memory of Bloody Sunday in Derry, Brian Conway illuminates how this event was "heard" differently by different actors and at different times, arguing that commemorations have their own historical pathways that are "reconstructed over time in response to shifting contexts in the remembering society and beyond" (2010: 150). He argues:

Some cultural objects may well be understood without being attentive to their historical pathways, but collective memory is not one of these. The study of collective memory compels us to pay attention to the historical construction of the past and the variable and often surprising paths memories take over time. Remembering and commemoration *have a dynamic past of their own* (emphasis added). Mnemonic life cycles are complex and shaped by human agents who mobilise political, social and cultural resources in support of their commemorative activity.

(Conway, 2010: 159)

Thinking this particular form of historical investigation of memory "not concerned with the past as such, but only with the past as it is remembered", Jan Assmann calls it mnemohistory (1997: 8–9). Mnemohistory is not about ascertaining the possible truth of traditions, but studying them as phenomena of collective memory. It does not look at determinants, but their effects; instead of actions remembered or commemorated, it examines the marks they have left; it does not focus on events but on their construction in time, their gradual disappearance and reappearance; and finally, instead of the past as it was, its constant re-exploitation, utilisation and manipulation, not the tradition itself but the way it was constituted and transmitted (Nora, 1992: 24; Tamm, 2008: 502).

Hence, this book does not examine the Second World War and its aftermath as an event but their construction in time. I am interested in the memory of this specific historical period because of its continued pertinence in Yugoslav and Serbian society. In analysing this, I am not concerned with proving that the historical narratives I am dealing with are incorrect. Memory as a historical source should be validated against objective evidence, but, "for a historian of memory, the 'truth' of a given memory lies not so much in its 'factuality' as in its 'actuality'" (Assmann, 1997: 9). Mnemohistory is about the continuous relevance of historical events, it "analyses the importance which a present ascribes to the past" (Assmann, 1997: 10). To adapt Jan Assmann's example, this mnemohistorical endeavour does not ask: Were the Chetniks antifascists? Was the postwar process of settling accounts with the issue of collaboration ideologically motivated? Instead, it poses the questions such as: Why did the Chetniks become central? Why were such statements articulated publicly and became dominant in Serbia only after the fall of Slobodan Milošević in 2000? Why is, for instance, the image of the Partisans based on the retributions at end of the war and in its immediate aftermath, and not on their victory?

The diachronic dimension of this book introduces the post-Yugoslav context and places contemporary memory politics in its historical setting. It sheds light on the historical development and transformations of the dominant narratives about the past, something that memory studies often overlook. Through the historical lens, I trace the continuities and ruptures concerning narratives, practices and actors of memory politics. Historicising the memory culture of post-Yugoslav Serbia is crucial for understanding the contemporary context.

Politicality of memory

During the *Historikerstreik* controversy in West Germany in the 1980s, historian Christian Meier used the term politics of memory (*Geschichtspolitik*) for the first time, as a pejorative name for politics interfering with historical interpretations. In the following years, memory politics developed into a theoretical and analytical concept in German scholarship (Troebst, 2013: 17). Politics of memory encompasses all discourses and acts where the interpretation of history is pursued for political purposes, as a public representation of the relevant past in the present (Schmid, 2008: 78). As such, politics of memory has three basic purposes: the establishment of traditions, the influence on collective identity and engendering the political and historical legitimacy (Schmid, 2008: 78).

As is clear from its very name, politicality is a central characteristic of memory politics. In the words of the Popular Memory Group, politics and history always intertwine, meaning that "all political activity is intrinsically a process of historical argument and definition, that all political agendas involve some construction of the past as well as future and that these processes go on every day" (1998: 78). Historical memory is always a political act (Wolfrum, 2010: 15). Similarly, Michael Bernhard and Jan Kubik argue that public remembering of the past is always a political process and, thus, politics of memory needs to be studied "as an integral part of the establishment of new collective identities and new principles of political legitimacy" (2014: 3). Memory politics is about interests, power and hegemony (Wolfrum, 2010: 19).

Besides politicality, another central characteristic of history and politics of memory is their public nature. When we talk about official policies of history, as Henry Rousso asserts, we are talking about public discourses and public actions, including commemorations (2011: 234). We do not focus on memory as a whole, but predominantly on what is public about it, what is seen, what is heard or at least what is visible and accessible to a large audience (Rousso, 2011: 234). The public stage and audience are necessary for public representations of history as opposed to private memory (Popular Memory Group, 1998: 76).

Memory politics always takes place in a context, in memory cultures. Memory cultures are common historical frames of reference that work on different levels, of nation-states, political groups or language and religious communities (Molden, 2011). It is an umbrella term that comprises of state-sponsored memory politics, non-state and civil society actors, families and experience-based memory of contemporaries (Cornelißen, 2003; Troebst, 2013: 29), similar to social frames of remembrance of Maurice Halbwachs (1992). The concept implies both historical and cultural variability when spoken in plural (Erll, 2011: 49).

Memory cultures as structural frames within which memory politics can take place are based on the common knowledge of history but they are not a homogeneous sphere. They involve different actors and changing power relations and encompass a variety of more specific memory cultures that struggle for the status of the shared and familiar history (Molden, 2009: 36). Within a wider memory culture of society, there are competing, coexisting and complementing memory subcultures of different social groups, that I also refer to as memory communities or milieus (Bergenthum, 2005: 128). No dominance or hegemony of memory is long-lasting or static (Molden, 2009: 39). Memory cultures, states and their institutions, ruling parties or coalitions or social and cultural groups are themselves heterogeneous and have to constantly renegotiate their views of the world and history in relation to changing political and economic challenges and transformations of society (Molden, 2009: 45).

State agency

While memory politics is intermixed with power, hegemony, legitimacy and identity-building, its agency is not only limited to the state. Nevertheless, the politics of memory often refers exclusively to the way "states, state governments, political parties and other elite groups have sought to encourage views of the past which serve their own ends", usually in relation to the acquisition, consolidation and extension of power (Berger and Niven, 2014a: 7). This is not surprising. As James Wertsch explains, states are naturally not the only entities that work on collective memory in the modern world, but they are "unrivalled in the power and resources they have devoted to this effort" (2002: 10). The modern state outweighs the struggles of different groups in purveying certain accounts of the past in resources, authority and restriction of competing efforts (Wertsch, 2002).

The state assumes the leading role in the processes of preserving and adapting war narratives, by putting

> in place a wide range of public institutions (war graves commissions, museums, libraries, archives, official histories) to maintain war memorials and transmit the narratives surrounding them. It may also assume responsibility for the welfare of the returned servicemen who fought in its name, and of the bereaved relatives of those who died.
> (Ashplant, Dawson and Roper, 2000: 25)

Because of that, the state-sponsored efforts still represent a valuable and natural arena for studying collective remembrance (Wertsch, 2002: 68). Other institutions generate certain views on the past too, like museums, galleries, record offices, media, publishers and NGOs, but historical constructions are most obviously public when linked to central state institutions (Popular Memory Group, 1998).

The constructions of the past that state actors purvey do not have to be false, but they are often one-sided and tendentious, generated and adapted to suit the interests of the present (Berger and Niven, 2014a: 7). Studying state-sponsored memory politics is not investigation of the authenticity of narratives and their correction. The question is not whether the communicated image of history is scientifically truthful (Meyer, 2008: 176). The question is "how and by whom, as well as through which means, with which intention and which effect past experiences are brought up and become politically relevant" (Meyer, 2008: 176).

The notions of official and dominant memory are particularly fruitful for understanding the dynamics of memory politics. Closely linked to the state efforts, official memory encompasses "those dominant or hegemonic narratives which underpin and help to organise the remembrance and commemoration of war at the level of the nation-state" (Ashplant, Dawson and Roper, 2000: 22). The official politics of memory takes place at all levels of society. Leaders or authorities in towns, ethnic communities or in educational, government and military bureaucracies share the interest "in social unity, the continuity of existing institutions and loyalty to the status quo", including reducing power of competing interests that might threaten their goals (Bodnar, 1992: 13).

Dominant memory, as the Popular Memory Group coined it, relates rather to "the power and pervasiveness of historical representations, their connections with dominant institutions and the part they play in winning consent and building alliances in the processes of formal politics" (1998: 76). This does not mean that everyone believes in the interpretation of the past that constitutes dominant memory, nor that every historical representation that reaches the public is dominant. It also does not have to represent the truth, but dominant representations can be most ideological or most obviously conforming to the "flattened stereotypes of myth" (Popular Memory Group, 1998: 76). Dominant memory are representations of the past that achieve centrality and luxuriate grandly, while others are marginalised, excluded or reworked.

Pluralities, struggles and layers

Multiple competing forces and interests exist in every mnemonic arena, "competing for hegemony of discourse and interpretative patterns" (Meyer, 2008: 176). Inspired by the Gramscian cultural hegemony theory, Berthold Molden thinks of memory politics as "determined by the relations of forces between hegemonic master narratives, defiant counter-memories and silent majorities whose historical experience is rarely articulated in the public" (2016: 125). The interactions between these different levels of memory politics are, in Gramscian terms, "a hegemonic process of ideological domination and resistance" (Ashplant, Dawson and Roper, 2000: 13). In addition to this struggle, many narratives and actors, often the majority in society, have passive and coexisting natures (Molden, 2014: 217).

Understanding of the sphere of memory politics as a struggle for hegemony does not imply a dichotomy, although the focus on either state or social agency still dominates the field. Addressing this issue, Bernhard and Kubik argue that "thinking of historical memory as a two-sided struggle can lead to oversimplification" (2014: 10). A diverse range of memory subcultures, communities and milieus articulate their memory through different social arenas, state agency, civil society or more informal groupings (Ashplant, Dawson and Roper, 2000: 16). They can shape national memory "only if they command the means to express their visions and if their vision meets with compatible social or political objectives and inclinations among other important social groups" (Rousso, 2011: 231).

The central activity that takes place at the heart of these struggles for public articulation and recognition is memory work. This book draws on Conway's multi-level concept of memory work that focuses on agency and understands that "memory is something that involves human effort – that is accomplished – over time and that can be 'successful' or not" (2010: 12). The idea of memory work closely relates to the theory of mnemonic hegemony, having a highly contested, fragmented and polarising nature at its core, as social groups engage in struggles over meanings and their ownership. Memory is "deeply implicated in power struggles and claims to legitimacy and domination" (Conway, 2010: 147). Embedded in social, political and institutional context, the success of memory work depends on its resonance and connection to these contexts (Conway, 2010: 149).

According to Conway, memory work operates at four levels: individual, small-group, social and institutional. The individual level is about private memory and involves activities individuals participate in to keep the past alive, including keeping photographs of the dead, attending commemorations or passing on stories (Conway, 2010: 5). The small-group level comes to being when these individuals come together to remember the past.

The small-group level of memory work is where memory choreographers appear, the actors who "are involved in creating and propagating commemorative discourses" (Conway, 2010: 6). They can be activists, families, community leaders, academics, journalists or politicians, the concept close to Michael Pollak's memory entrepreneurs as actors who create common references and make sure they are respected (Jouhanneau, 2013: 25). For Bernhard and Kubik (2014), these actors are mnemonic agents: they can be guided by moral or legal imperatives and political purposes.

Normally leading to a public act of commemoration, the small-group memory work transforms into the social level. As Conway shows in the example of Bloody Sunday, the social memory work usually structures around the social calendar and "it brings struggles between different social groups over issues of ownership of the past to the fore" (2010: 146). National memory involves structures revolving around anniversaries of historical events and their commemorations that represent the "mental historical calendar of the nation" (Berger and Niven, 2014b: 146). Each commemorative act or ritual reproduces a commemorative narrative that reconstructs a specific fragment of the past. A set of commemorations together form the master commemorative narrative that official politics of memory builds upon and that aims at structuring collective memory. In her study on politics of memory and construction of national myths in Israel, Yael Zerubavel defines these master commemorative narratives as "a broader view on history, a basic 'story line' that is usually constructed and provides the group members with a general notion of their shared past" (1995: 6). Individual commemorative acts contribute to the formation of the master commemorative narrative which is why they should be studied within the wider context of the narrative they reproduce, and not in isolation.

State institutions and organisations have a crucial role in doing memory work at the institutional level. This is the sphere of official politics of memory. All four levels of Conway's analysis are based on two key properties: they are mutually embedded, and they exercise different levels of authority and power. First, memory work at one level tends to be oriented to memory work at another level, like the social memory work is usually oriented towards changing or affirming state-sanctioned memory of the past (Conway, 2010: 7). Moreover, there is a hierarchy of memory work where the state level "tends to exert the strongest claim on power and authority" (Conway, 2010: 7). Moving from micro to macro levels of this hierarchy means moving from multiple individual memories to increasingly abstract and homogenised memory work at the social and institutional level. Already within the social-level memory work, the issue of fragmented and plural individual memories is overcome by the suppression and simplifications of some of them in order to construct a shared group memory (Conway, 2010: 8).

Post-Yugoslav Serbia as a case study

This book engages with the question of how politics of memory works. Exploring memory politics inevitably encompasses the state-sanctioned or institutional level of memory work. Moreover, to fully grasp the dynamics of memory politics, as opposed to creating a mere catalogue of state institutions and their work, I consider the multiple layers of memory and their interactions and overlaps. By examining the different layers of politics of memory, their interrelations and actors, this book moves away from the strictly state-oriented focus, understanding that it is not states who remember but "individuals do, in association with other people" (Winter, 2006: 4). At the same time, the study avoids the common binary perspective that only examines state actors and historical narratives they promote on the one hand, and the opposition to them on the other, replacing it with "a more open, polyphonic view on mnemonic agency" (Molden, 2016: 130).

This book demonstrates that national frameworks still retain their relevance, even in the context of the transcultural and global turn in social sciences, humanities and memory studies. It takes an empirical approach to memory cultures, politics and practices that fragments the often homogenising global, European, East European and postsocialist perspectives. Rather than proclaiming the uniqueness of my case study, I understand it as a starting point. The focus on specificities such as the Serbian case could encourage more valid comparisons with consequential implications for the general understanding of politics of memory. Finally, a historically contextualised and in-depth analysis of memory politics could be applied more broadly, not only to studying memory of the Second World War and state socialism in postsocialist countries and Europe.

Contemporality and troubles with materials

The book relies on a wide range of written sources that help deconstruct the dynamics of memory politics, identify its actors and grasp decision-making processes. The process of locating the materials and gaining access to them when studying the most recent past is significantly different than doing archival research. This difference concerns the nature of the sources, different paths towards accessing them and the issues that arise. The state of the sources, as I will illustrate below, is a rationale for the turn to the empirical and it increases the significance of media reports in the endeavour to reconstruct and fully understand an ongoing phenomenon such as memory politics in contemporary Serbia.

Since 2000, the state has introduced numerous legal and institutional mechanisms directed at the reinterpretation of the Second World War and socialist Yugoslavia. The National Assembly of the Republic of Serbia has discussed each of these measures and complete transcripts are available online through the Open Parliament (*Otvoreni parlament*) project.[1] The project provides all parliamentary transcripts and detailed information on all members of the parliament since 2000, including their party affiliation, statements, proposals and the history of voting. Furthermore, *Službeni glasnik Republike Srbije* is a state publication of

> acts of the National Assembly, the president of the Republic, the government, ministries, and special organisations of the Constitutional Court, the High Court Council, the State Council of Prosecutors, courts and public prosecutor offices, the National Bank of Serbia, [...] and corrections of acts.
> (Službeni glasnik Republike Srbije, n.d.)

Most state institutions have the information regarding their work available on their websites. Hence, a large portion of materials of state affiliation in Serbia is very easily accessible for researchers, in a similar manner as in most countries.

The materials that are not public have to be requested directly from the institutions, according to the Law on Free Access to Information of Public Interest (*Službeni glasnik Republike Srbije*, 2004). This was the way I was able to receive, for instance, the 2009 and 2013 State Program for Commemorating the Anniversaries of Historic Events of the Serbian Liberation Wars, the documents describing "the significant anniversaries, commemorative days, national and religious holidays, as well as important Serbian personalities, together with a brief description of why specific days are commemorated and exhaustive explanations of the protocol for each of them" (David, 2014: 477).

In order to gain access to materials related to rehabilitation processes, one must request transcripts of hearings and court decisions for each case individually. Two issues arose in this process. The first problem is that, unlike in the archives where the documents are ready for the researcher on the following day at the latest, requesting materials from public institutions can be a very long-lasting process. I submitted the request to receive access to the entire case of Dragoljub Mihailović one week after he had been rehabilitated in May 2015. After three resubmissions, the paperwork involving choosing the form in which I would like to receive the requested materials and fee payments, I received printed copies of the documents by post in January 2016.

28 *Exploring politics of memory*

Another issue is that, of course, the access to documents is only possible after a court process ends, which also relates to the wider issues of the mechanism of judicial rehabilitation in Serbia. Namely, the 2011 Rehabilitation Law abandoned the use of non-adversarial proceedings and introduced the prosecutor's office as the other side in the courtroom, additional to the plaintiffs. According to the 2006 law, the proceedings were non-contentious and involved only plaintiffs, and there was no right of complaint, meaning that a process would end with the court decision. This was the case with Dragoljub Mihailović. My observation of the rehabilitation practice according to the 2011 law illuminates that the prosecutors remain mostly passive throughout processes and their activity is based on filing complaints against the court decision at the end. These complaints that can repeat numerous times, can complicate and prolong rehabilitation processes. Prosecutors file complaints to the Court of Appeal in Belgrade, which usually sends the case back to the original court. During this circle, the documents remain inaccessible for researchers.

The inaccessibility of the judicial rehabilitation materials in the cases of high relevance such as Milan Nedić and Nikola Kalabić represent one of the factors that pushed this research in the empirical sphere, especially as it became clear that no documents would be made available before the completion of this manuscript. Throughout 2016 and 2017, I attended several court hearings in the process of rehabilitation of Milan Nedić, taking notes of the testimonies and observing the atmosphere and the actors in the courtroom. The judge at the High Court in Belgrade had to allow my presence, and, because of the great public interest in the case and the small courtroom, this had to be negotiated prior to every proceeding. The plaintiffs' lawyer, Zoran Živanović, who I had interviewed earlier, played a decisive role a few times by lobbying to the judge.[2] Attending the court hearings enabled me to observe the dynamics of protests in front of the court building, too. Moreover, it helped me identify the journalists and media who regularly attend the hearings, whose reports on the case I could follow when I was not able to attend it myself. Reliance on media reports is unavoidable in these cases. In this way, in the case of Nikola Kalabić, whose process has taken place in Valjevo, I was not able to attend hearings, but I relied on the media reports who paraphrased and cited witness testimonies as well as court decisions.

The experience of officially accessing the materials of rehabilitation processes illuminates how inconsistent the court practice in this field in Serbia is, which I will further discuss in the legislation chapter. One of those inconsistencies is the approach to the privacy of personal information and I will present three cases here to illustrate it. In the aforementioned documents from the case of Dragoljub Mihailović, all information was visible except the redacted names of the witnesses. The redacted names were, however, easily readable, due to their redaction by hand, using a black marker pen that was transparent. The witnesses of this case had also been familiar to me as the media reports while the case was ongoing made it clear who they were, and some had even spoken publicly about their testimonies. Another example is the rehabilitation of Zorka Roknić Kragić in the court in Pančevo, in the Serbian province of Vojvodina.[3]

The High Court in Pančevo sent me the court decision with so much redacted information that it was impossible to read or understand anything at all.

A completely different example is that of Bogdan Lončar and Milenko Braković, rehabilitated at the Regional Court in Šabac in 2008.[4] After numerous unsuccessful attempts to reach the court archive and request the documents, Gojko Lazarev gave them to me, the judge who had presided over the case and the court president at the time of our meeting in 2016. After we discussed this rehabilitation case and my brief remark that I had not managed to obtain the documents, Lazarev called the court archive and had them copy the entire case and bring it to me, with all materials included. As opposed to this rather informal experience, the archive of the court in Niš in the south of Serbia has never replied to any of my enquiries after the court administration pointed me in their direction for the 2009 rehabilitation case of Dragiša Cvetković, the prime minister of the Kingdom of Yugoslavia who had signed the Tripartite Pact.

There are always public bodies that do not generate any sources, or the sources exist but are entirely unavailable for researchers. Two examples of the scarce documentation are the governmental Commission for State Holidays (*Komisija za državne praznike*) established in 2001, and the Board for Fostering Traditions of the Liberation Wars of Serbia (*Odbor za negovanje tradicija oslobodilačkih ratova Srbije*), composed of eight representatives of several ministries. I was particularly interested in potential transcripts of meetings and the process of decision-making regarding official holidays and commemorations, but I did not manage to obtain them.

More important for this thesis is the State Commission for the Determination of Circumstances of the Execution of Dragoljub Mihailović (*Državna komisija za utvrđivanje okolnosti pogubljenja Dragoljuba Mihailovića*) that does not even have an official statute or any other document and which had not left any trace on any institutional or ministerial internet page. In order to reconstruct its work, I had to turn to media reports of the Commission's quest for Mihailović's remains, while resorting to interviewing the Commission members, predominantly historians. Finally, as a group of historians acts as memory choreographers in state-sanctioned bodies and rehabilitation processes, their publications that refer to these contexts and their own memory work are taken into consideration as primary sources, rather than secondary literature.

To conclude, the abovementioned examples illustrate the diversity in the availability of sources and the difficulties I encountered during the research in Serbia. The problem of inaccessibility or the lack of documents vital for the reconstruction of the work of public bodies and memory work processes was overcome by turning to media and, more importantly, by interviewing the most essential actors of memory politics – memory choreographers.

Oral sources

Doing research on such a recent period, as this book explores, is both a blessing and a curse. On the one hand, when the research subject is an ongoing phenomenon, it can be very challenging to draw a line. On the other hand, the contemporary nature of the subject enables the researcher to complement the research with the use of oral sources, as all actors of the events and processes examined are alive. For this book, I went beyond the written sources and engaged in interviews and participant observation. Such an approach gives face to politics of memory – its actors. People are always behind the mnemonic agency and its structures and the focus on state and other institutions and mechanisms of memory politics often neglects them.

As Paul Thompson says, oral evidence can be "evaluated, counted, compared and cited along with the other material. It is no more difficult, no easier. But in some ways, it is a different kind of experience" (2000: 271). Conducting interviews with memory choreographers of multiple levels of memory work in Serbia was certainly a different kind of experience for me. The interviews were semi-structured, somewhere between "rigidly structured so-called box-ticking 'questionnaires' and free flowing dialogue" (Thompson, 2000: 225). Politicians, judges, historians, members of political parties and the parliament and memory activists, including descendants and relatives of people considered victims of communism, made up the interview sample. None of the informants had a direct experience of the Second World War or its aftermath, but some had indirect experience either through their family or their profession.

Many interviewees are historians. First, I am interested in historians who act as agents of state-sanctioned memory politics and at the small-group and social level of memory work: those critical of their interpretation of history usually refer to them as "revisionists". The conversations with them were like those of other memory choreographers, but I asked them to also specifically address their motivation and their double role of a professional historian and a mnemonic agent. I was interested in the political motivations behind their activism and their networks. In the case of historians not acting as agents of official memory politics, my sampling was also specific and included historians who publicly criticised the reinterpretation. These interviews could be labelled expert interviews, directed towards people whose knowledge comes through their profession or education, who have power in creating knowledge among non-experts, going after their specific knowledge (Bogner, Littig and Menz, 2009).

In addition to historians, the book is based on the series of interviews with mnemonic agents of all Conway's levels of memory work, from individual to institutional. The book moves beyond the prevalent idea that state-sanctioned politics of memory in Serbia, as revisionist as it may be, is simply imposed from above and does not resonate in society. Moreover, state-sanctioned memory politics is not an abstract sphere but a field of very concrete memory work and its agents. Hence, I studied it as such. As discussed above, this does not imply a dichotomy of state agency and civil society, but a sense for the plurality of memory politics and its multi-layered nature.

Although I have also conducted interviews with liberal and leftist memory activists who reject and oppose the official memory, I am more interested in the completely neglected sphere of anti-communist memory work. This is where public acts of commemorations of the defeated side of the Second World War happen, this is the memory work that commemorates victims of communism and that generates memorials. This focus also goes against the reductive romantic dichotomy of leftist and liberal civil society that stands opposed to the state. These actors will be referred to as anti-communist throughout the thesis, because, although they come from across the spectrum of conservative and right-wing politics, anti-communism is what they all have in common. Because of their views, the role in memory politics and the support for anti-communist memory work on a small-group and social level, revisionist historians belong to this sphere.

Even though the book is not an oral history project and uses the oral evidence alongside other sources, they form a vital part of its foundations. If we understand oral history as both the practice, the process of conducting and recording interviews, and the output, the result of that research process, this book employs oral history only as a research methodology (Abrams, 2010: 1). Examining collective memory with the help of oral history is not a novel approach. As James Mark explains, oral history

> allows the historian to examine the impact of new public memories on individuals' conceptions of their own lives, and the way in which these interacted with, and in some places replaced, the ways that they had learnt to think about their lives under the previous regime.
> (Mark, 2010: xxii)

Many of my interviewees are memory activists whose private memories relate to a public memory of communism and are the main driving force behind their activism. The life stories and the adaption of biographies to suit the present is not my main objective, although the informants usually seek to "forget, or downplay, their stories of social mobility and professional success" (Mark, 2010: xxv). Many of my informants compose their memory of communism by repressing memories that do not accord with their present identity and public memory, such as gaining an education and a career and earning their pensions in socialist Yugoslavia. They reconceptualise themselves "as victims of communist criminality who had been liberated by democracy" (Mark, 2010: xx). I am also interested in illuminating how they remember the more recent past and how they experienced the important events such as the overthrow of Slobodan Milošević and the dissolution of Yugoslavia.

Interested in history from below, oral and social historians have traditionally been oriented towards the marginalised groups, often empowering them through giving them a voice. They have paid less attention to "to the life stories of ordinary people whose political agendas they find unsavoury, dangerous or deliberately deceptive" (Blee, 1993: 596). As Kathleen Blee demonstrates, oral historians have usually emphasised caution, distance and objectivity when interviewing members of elites, and egalitarianism, reciprocity and authenticity in interviews with people outside elites (1993: 597). She criticises the epistemological dichotomy in oral history that reflects romantic assumptions about the subjects of history from the bottom-up, difficult to defend "when studying ordinary people who are active in the politics of intolerance, bigotry or hatred" (1993: 597). She draws her conclusions from working on the remarkable study of the women of the Ku Klux Klan (Blee, 2008). Although my informants are far from being members of such a militant and racist group that persecutes others, my approach nevertheless implies the immersion in the sphere of often far-right politics.

I expected my informants to be reluctant to talk to me. This did not happen, and I had no difficulty arranging the meetings. There was an assumption of commonality from their side, as if they found it impossible to imagine that I might be critical towards their views. Even controversial topics and questions and our discussions beyond the interviews did not seem to raise their awareness of my potential critical standpoint. My age and gender were important factors too, so many of my informants probably perceived themselves in positions of authority and seniority and, for instance, some referred to me as "child" and "our child" on several occasions. People I expected to dislike, because of their views, biographies or positions, were interesting, intelligent, well-informed and very pleasant to talk to (Blee, 1993: 604–605). I was also unwilling to do anything that could ruin these dynamics.

In the explorations of memory politics for this book, I did get "the seat of my pants dirty in real research" (Park in Brewer, 2000: 13).[5] Relying on the combination of participant observation, interviews and a collection of predominantly written sources, the approach of this book could be broadly termed ethnographic or, more precisely, having ethnographic leanings. In addition to attending court proceedings, I attended commemorations, public discussions, lectures and protests and I engaged in conversations and various encounters while doing participant observation. This and interactions with informants generated more primary sources that might otherwise be overlooked, including flyers, brochures, museum and exhibition catalogues, books, photos and copies of diverse documents the informants insisted I take into consideration. Finally, considering the semi-official and unofficial nature of most memorials to the defeated and victims of communism that I wanted to map, a lot of time during the fieldwork was dedicated to what could best be described as monument hunting.

Notes

1 Four non-governmental organisations launched the project in 2012, funded by the British Embassy in Serbia. The website includes the stenographic notes since 2000.
2 The names of my informants are not anonymised in this book because they all signed a consent form allowing their name and position to be published in my research.
3 In 1942, she reported the communist resistance meeting to the German authorities, leading to the death of Žarko Zrenjanin, the secretary of the Provincial Committee of the Communist Party of Yugoslavia. While he was at the meeting, preparing to travel to Bosnia to attend the first assembly of the Antifascist Council for the National Liberation of Yugoslavia in Bihać, the police and German occupation forces surrounded the house

in the village of Pavliš. Roknić Kragić knew about the location because of her involvement in the resistance movement. After her capture and hearing, she was executed after the liberation in 1944.
4 Bogdan Lončar and Milenko Braković were police gendarmes shot by the Partisan Živorad Jovanović Španac on 7 July 1941, the day later celebrated as the Day of the Uprising of People of Serbia.
5 Robert Park in the speech to the students at the University of Chicago in the 1920s.

Bibliography

Abrams, L. (2010) *Oral History Theory*. London: Routledge.
Ashplant, T. G., Dawson, G. and Roper, M. (eds) (2000) *The Politics of War Memory and Commemoration*. London: Routledge.
Assmann, J. (1997) *Moses the Egyptian: The Memory of Egypt in Western Monotheism*. Cambridge, MA: Harvard University Press.
Bergenthum, H. (2005) "Geschichtswissenschaft und Erinnerungskulturen. Bemerkungen zur neueren Theoriedebatte". In Oesterle, G (ed.), *Erinnerung, Gedächtnis, Wissen. Studien zur kulturwissenschaftlichen Gedächtnisforschung. Formen der Erinnerung 26*. Göttingen: Vandenhoeck & Ruprecht, pp. 121–162.
Berger, S. and Niven, B. (2014a) "Introduction". In Berger, S. and Niven, B. (eds), *Writing the History of Memory*. London: Bloomsbury Academic, pp. 1–24.
Berger, S. and Niven, B. (2014b) "Writing the History of National Memory". In Berger, S. and Niven, B. (eds), *Writing the History of Memory*. London: Bloomsbury Academic, pp. 135–157.
Bernhard, M. H. and Kubik, J. (eds) (2014) *Twenty Years After Communism: The Politics of Memory and Commemoration*. Oxford: Oxford University Press.
Blee, K. M. (1993) "Evidence, Empathy, and Ethics: Lessons from Oral Histories of the Klan", *The Journal of American History* 80(2): 596–606. DOI: 10.2307/2079873.
Blee, K. M. (2008) *Women of the Klan: Racism and Gender in the 1920s*. Oakland: University of California Press.
Bodnar, J. E. (1992) *Remaking America: Public Memory, Commemoration, and Patriotism in the Twentieth Century*. Princeton: Princeton University Press.
Bogner, A., Littig, B. and Menz, W. (eds) (2009) *Interviewing Experts*. Basingstoke, New York: Palgrave Macmillan.
Brewer, J. D. (2000) *Ethnography*. Buckingham, Philadelphia: Open University Press.
Burke, P. (1997) "History as Social Memory". In *Varieties of Cultural History*. Ithaca: Cornell University Press, pp. 43–60.
Conway, B. (2010) *Commemoration and Bloody Sunday: Pathways of Memory*. Palgrave Macmillan Memory Studies. Basingstoke: Palgrave Macmillan UK.
Cornelißen, C. (2003) "Was heißt Erinnerungskultur? Bergriff – Methoden – Perspektiven", *Geschichte in Wissenschaft und Unterricht* 54: 548–563.
David, L. (2014) "Impression Management of a Contested Past: Serbia's Evolving National Calendar", *Memory Studies* 7(4): 472–483. DOI: org/10.1177/1750698014537670.
Erll, A. (2011) *Memory in Culture*. Basingstoke: Palgrave Macmillan.
Halbwachs, M. (1992) *On Collective Memory*. Chicago: University of Chicago Press.
Jouhanneau, C. (2013) "Would-be Guardians of Memory: An Association of Camp Inmates of the 1992–95 Bosnian War under Ethnographic Scrutiny". In Mink, G. and Neumayer, L. (eds), *History, Memory and Politics in Central and Eastern Europe: Memory Games*. Basingstoke: Palgrave Macmillan, pp. 23–38.

Mark, J. (2010) *The Unfinished Revolution: Making Sense of the Communist Past in Central-Eastern Europe*. New Haven: Yale University Press.
Meyer, E. (2008) "Memory and Politics". In Erll, A. and Nünning, A. (eds), *Cultural Memory Studies. An International and Interdisciplinary Handbook*. Berlin: De Gruyter, pp. 173–181.
Molden, B. (2009) "Mnemohegemonics. Geschichtspolitik und Erinnerungskultur im Ringen um Hegemonie". In Molden, B. and Mayer, D. (eds), *Vielstimmige Vergangenheiten. Geschichtspolitik in Lateinamerika*. Vienna: LIT, pp. 31–56.
Molden, B. (2011) "Geschichtspolitik als politisches Handlungsfeld". Available at: www.linksnet.de/index.php/artikel/26350 (accessed 19 May 2018).
Molden, B. (2014) "Mnemonic Hegemony? The Power Relations of Contemporary European Memory". In Boyer, J. W. and Molden, B. (eds), *EUtROPEs: The Paradox of European Empire*. Parisian Notebooks. Paris: University of Chicago Center, pp. 104–131.
Molden, B. (2016) "Resistant Pasts Versus Mnemonic Hegemony: On the Power Relations of Collective Memory", *Memory Studies* 9(2): 125–142. DOI: org/10.1177/1750698015596014.
Nora, P. (1992) "Comment écrire l'histoire de France?" In Nora, P. (ed.), *Les Lieux de Mémoire, III: Les France, 1: Conflits et Partages*. Paris: Gallimard, pp. 11–32.
Popular Memory Group (1998) "Popular Memory: Theory, Politics, Method". In Perks, R. and Thomson, A. (eds), *The Oral History Reader*. London: Routledge, pp. 75–87.
Rousso, H. (2011) "The History of Memory. Brief Reflections on an Overloaded Field". In Blaive, M., Gerbel, C. and Lindenberger, T. (eds), *Clashes in European Memory: The Case of Communist Repression and the Holocaust*. Innsbruck: Studien Verlag, pp. 231–239.
Schmid, H. (2008) "Konstruktion, Bedeutung, Macht. Zum kulturwissenschaftlichen Profil einer Analyse von Geschichtspolitik". In Heinrich, H.-A. and Kohlstruck, M. (eds), *Geschichtspolitik und sozialwissenschaftliche Theorie*. Stuttgart: Steiner, pp. 75–98.
Službeni glasnik Republike Srbije (2004) Zakon o slobodnom pristupu informacijama od javnog značaja.
Tamm, M. (2008) "History as Cultural Memory: Mnemohistory and the Construction of the Estonian Nation", *Journal of Baltic Studies* 39(4): 499–516. DOI: org/10.1080/01629770802468865.
Thompson, P. (2000) *Voice of the Past: Oral History*. 3rd edn. Oxford: Oxford University Press.
Troebst, S. (2013) "Geschichtspolitik. Politikfeld, Analyserahmen, Streitobjekt". In François, E., Kończal, K., Traba, R. and Troebst, S. (eds), *Geschichtspolitik in Europa seit 1989. Deutschland, Frankreich und Polen im internationalen Vergleich*. Göttingen: Wallstein Verlag, pp. 15–37.
Wertsch, J. V. (2002) *Voices of Collective Remembering*. Cambridge: Cambridge University Press.
Winter, J. M. (2006) *Remembering War: The Great War Between Memory and History in the Twentieth Century*. New Haven: Yale University Press.
Wolfrum, E. (2010) "Erinnerungskultur und Geschichtspolitik als Forschungsfelder". In Scheunemann, J. (ed.), *Reformation und Bauernkrieg. Erinnerungskultur und Geschichtspolitik im geteilten Deutschland*. Leipzig: Evangelische Verlagsanstalt.
Zerubavel, Y. (1995) *Recovered Roots: Collective Memory and the Making of Israeli National Tradition*. Chicago: University of Chicago Press.

3 Yugoslav memory culture and its downfall

> The unitary Kingdom of Serbs, Croats and Slovenes, renamed Yugoslavia in 1929, was destroyed and dismembered during World War II, only to re-emerge as a socialist federation and with some territorial gains in the west in its aftermath. It, too, broken up in a series of wars, beginning in 1991, only to continue in its final form in the alliance of Serbia and Montenegro, also under the name of Yugoslavia from 1992 to 2003 when it changed its name to Serbia and Montenegro, which then continued until 2006 when these two republics, in an agreed separation, became independent states.
>
> (Velikonja, 2017: 517)

Socialist Yugoslavia was born in the territory of the Kingdom of Yugoslavia and on the ruins of its socio-political and economic structure during the Second World War (Milošević, 2017: 353). The Partisans, under the leadership of the Communist Party of Yugoslavia, laid the foundations for Yugoslav state socialism already during the war by establishing political and state structures in the territories they liberated. After their victory in 1945, one of the immediate concerns of the new regime was the wartime enemies and the need to settle accounts with the issue of collaboration. Settling accounts took the form of extrajudicial executions and post-war trials. In this sense, Yugoslavia reflected the wider European context at the time where the initial phase of retributive violence prior to and immediately after the end of the war led to institutionalisation through tribunals and commissions (Judt, 2005: 41–62). The political and military forces who collaborated with the Axis occupation and fought against or persecuted the Partisans and their supporters faced punishment, such as Milan Nedić authorities in Serbia and the Ustasha regime in Croatia. As many managed to flee Yugoslavia, courts could only sentence them in absentia. Confiscation of property and declaration as a people's enemy (*narodni neprijatelj*) usually followed these measures.[1]

The period of consolidation of state socialism involved the elimination of political opposition, expropriation of land of Germans, large landowners and the Church and its redistribution and mandatory contributions from peasants. Nevertheless, the Party had wide popular support (Baker, 2015: 19). The support thrived among peasants from Herzegovina and Krajina who were resettled on the more fertile land in Slavonija and Vojvodina formerly owned by Germans; among women, who received the vote; among those who saw this Yugoslavia's multi-ethnic, federal structure as a guarantee against repression by Serbs or Croats.

(Baker, 2015: 19)

For many people, Yugoslav state socialism was a promise of emancipation and social mobility. For many Partisans, brotherhood and unity were more than ideological foundations of the new state, they represented their wartime experience (Karge, 2010: 24). The Partisans also acquired massive popular support across Yugoslavia during the war. Furthermore, the period of repression and implementation of the Soviet political and economic model did not last long. After the 1948 split with the Soviet Union, Yugoslavia oriented towards the development of the workers' self-management framework, becoming increasingly liberal and decentralised.

The People's Liberation War and so-called Partisan myth dominated the Yugoslav memory culture and built the foundations of the regime's legitimacy, serving this purpose beyond the country disintegration in the early 1990s and until the fall of Slobodan Milošević in 2000. However, the ethnonationalist lens that highlighted the parallel victimhood and heroism of the Serbian nation increasingly influenced the vision of the Second World War from the 1980s. During the rule of Milošević, these narratives became the official discourse in Serbia. The Yugoslav politics of balance that promoted equal participation of all Yugoslav people in both resistance and collaboration and ethnic neutralisation of the war experience did not fit the context of the state dissolution.

While the official interpretation of the Second World War did not drastically change between 1945 and the dissolution of Yugoslavia, the Yugoslav memory culture was not a monolithic sphere. People interacted diversely with the official mnemonic frameworks. Instead of merely complying with or rejecting the official narratives and practices, people organised grassroots commemorative initiatives, appropriated and transformed the official narratives, they were indifferent, and they also deliberately neglected or even destroyed Partisan memorials. The institutional level of memory work did not differentiate between different Partisans' enemies, meaning that the Chetniks represented the Serbian counterpart to the Ustasha, blamed for causing immense suffering to the Serbian people and obstructing the Partisans' struggle. Vernacular memory often revolved around families mourning the members of these forces or their commemorations, sometimes including memorials and plaques.

Yugoslav war memory

The Second World War and socialist Yugoslavia were inseparably linked to each other (Sundhaussen, 2004: 374). The war constituted both the founding event and myth for Yugoslav state socialism. The dominant war memory centred on the People's Liberation War (*Narodnooslobodilačka borba*, NOB), how the Partisans' struggle against the occupiers and their domestic helpers was officially termed, and victims of fascism. Other aspects of the war, such as inner Yugoslav conflicts, the Holocaust or the Roma genocide, were not relevant for the hegemonic discourses. Instead, the official war narrative encompassed three elements: the people's liberation, the parallel socialist revolution and amalgamation of the former into a single historical process leading to the brotherhood and unity of the Yugoslav people (Karge, 2010: 83). The war memory served a legitimising purpose for the regime and entailed an important integrating function because of its emphasis on the common struggle of all Yugoslav people against the occupation and domestic traitors. The authentic nature of the People's Liberation War as the Yugoslavs themselves liberated the country, not the Red Army, was a strong emancipatory factor (Koren, 2012: 509).

The central task of war memory was the creation of "a cult of heroism around those considered to have given their lives for the liberation and creation of the 'socialist homeland'" (Bergholz, 2010: 2). These were the Partisans, with the accent on their political, military and moral superiority over all other actors and ignoring all other options but the Communist Party as historically illegitimate (Höpken, 1999b: 197–198). The Yugoslav war interpretation was fragmented through the exclusion of a variety of war experiences and simplification of the very complex nature of the Second World War in Yugoslavia. Such selective war memory was not unique to Yugoslavia or state socialism. It spread across Europe's East and West in the postwar decades, with governments promoting a very selective memory of the wartime events and suffering, aiming at legitimating and stabilising their political orders and promoting the ideas of national recovery and cohesion (Bergholz, 2010: 3).

The Second World War in Yugoslavia was also a complex and multi-sided civil war. Its remembrance in the common Yugoslav state had to involve ethnic neutralisation. Thus, domestic traitors came from all Yugoslav nations and were the only side responsible for the war, crimes and terror, while the Partisans and the Party stood opposed to them as a transethnic symbol of resistance of all the Yugoslav people (Höpken, 2006: 408). The memory of a joint struggle and suffering fostered consensus in a society fragmented along ethnic, cultural and religious lines and burdened by the unfavourable experiences of living together in one state in the past (Höpken, 1999b: 197). For this reason, the notion of multinational and transnational solidarity constituted the basis of war memorialisation – with the focus on the Partisan resistance – and built the core of new socialist Yugoslavia (Kirn, 2016: 315).

Main mnemonic agency

The Partisan veteran association, SUBNOR, was the main actor responsible for the preservation of the memory of the war and revolution. With one million members and Josip Broz Tito as the president, SUBNOR was more than a veteran association, serving as "memory watchdog" (Höpken, 1999b: 197). Its simultaneous role as memory maker and veteran association was logical, taking into consideration that the hierarchy of the Yugoslav war memory placed the fallen Partisan fighters at the highest position, next to the veterans who survived the war and above the victims of fascism, understood as all people who died without a weapon in their hands (Karge, 2010: 46–47). SUBNOR had three main tasks. The most important was social care for war invalids, orphans, families of fallen Partisans and the Partisans who had survived the war. The further and equally important task was memory work. SUBNOR took care of preservation and transmission of the legacies of the People's Liberation War, including maintenance and building of graves and memorials and collection of historical documents and testimonies. Finally, the association was the official representative of the People's Liberation War abroad (Karge, 2010: 35). While SUBNOR operated at all levels of society, memory work primarily took place at the local level, as local communities organised and financed commemorative practices and related activities, taking care of graves, re-enactment marches, celebrations and trips to historical sites.

The Yugoslav memoryscape was abundant in monuments and memorial complexes. Their building started at the end of the war and, by 1961, the veteran association had built 14,402 monuments and cemeteries, which accounts for three per day during the 16-year period (Bergholz, 2006: 79–80). The revealing of graves or memorials was always a public event. SUBNOR organised it and thousands of people attended, including local veterans and officials, families of the fallen soldiers and pioneers, who actively participated in the program (Bergholz, 2006: 81). Schools, streets, public institutions and factories commemorated the fallen Partisans by carrying their names. More than 1,300 Partisan fighters were decorated with the Orders of the People's Hero, many posthumously.

In addition to SUBNOR, the federal authorities established the Board for Marking and Care of Historical Sites from the People's Liberation War in 1952 as an all-Yugoslav mnemonic agent based in Belgrade.[2] In Heike Karge's words, it was a Yugoslav mnemonic think tank (2010: 109). Its tasks involved the organisation, commemoration and preservation of memory of important events and dates of the Second World War, and construction and renovation of monuments and plaques (Horvatinčić, 2011: 98–99). Not fulfilling its large-scale memorial plans, the Board dissolved in the early 1960s, as the Party abandoned the all-Yugoslav perspective in memory politics, reflecting its move away from the development of the Yugoslav patriotic consciousness and identity (Karge, 2010: 109).

The Partisan myth that was at the core of the Yugoslav memory culture did not go through significant transformations during the Yugoslav period. There are, however, two phases in the history of memory politics that SUBNOR's work mirrored. During the postwar first decades, memory work was aimed at the war generation and involved an enormous volume of memorials built in the 1950s. From the 1960s on, the focus gradually shifted to the generations without a direct war experience and to the use of diverse media forms and popular culture. The modernisation of war memory for younger generations would transform the content of the official war memory into a truly collective cultural memory of the Yugoslav society (Karge, 2010: 43, 76). The popularisation efforts encompassed the distribution of memory through media forms and transformation of sites of memory into popular tourist places by adding restaurants, hotels and parking places (Horvatinčić, 2011: 102).

The split with the Soviet Union in 1948 also affected the Yugoslav war narratives. During the first postwar years, the framing of the Partisan struggle and victory was within the broader discourses of alliance of the Yugoslav and Soviet people, emphasising the leading position of the Soviet Union in the international communist movement and the importance of the Red Army in the victory against fascism. After the split, the emphasis shifted entirely to the Yugoslav Partisans as the carriers of the revolutionary transformation of society (Manojlović Pintar, 2010: 547). From this moment, the People's Liberation War would also serve "as an autochthonous socialist revolution, as worthy as that of the Soviet Union" (Sindbæk, 2012: 71).

The Day of the Uprising

Many Second World War events became official holidays celebrated across the country. One of them was 4 July, the day of the Communist Party leadership meeting in Belgrade in 1941 that formally decided on the organised uprising against the occupation. As a federal holiday entitled the Fighter's Day (*Dan borca*), 4 July symbolised the beginning of the uprising in Yugoslavia. Each Yugoslav republic had an additional day commemorating their own uprisings. Bosnia and Herzegovina and Croatia shared 27 July as the Day of the Uprising, commemorating the revolt of the Serb population against the Ustasha in Krajina, the bordering region between Bosnia and Croatia (Roksandić, 1995: 259).[3] Celebrating the beginning of the struggle against the occupation had a function of returning to the roots and events through which the Yugoslav communists "realised their historical role" and the people of Yugoslavia, by taking part in the uprising, acknowledged the communists as "the guarantee of a 'happier' future" (Roksandić, 1995: 257).

The Day of the Uprising of the People of Serbia (*Dan ustanka naroda Srbije*) was 7 July. On this day in 1941, a group of Partisans arrived in Bela Crkva in western Serbia to organise a political rally in front of the locals who had assembled for the annual village fair. Two Serbian gendarmes, Bogdan Lončar and Milenko Braković, came to disperse the gathering. Someone warned the Partisans who returned and confronted the gendarmes, with Živorad Jovanović Španac shooting and killing them both. The Presidency of the People's Assembly of Serbia declared this as the Day of the Uprising in July 1945, posthumously decorating Španac, who had also been a Spanish Civil War volunteer, as the People's Hero.

Everything about the commemorations of 7 July mirrored the Yugoslav memory politics. A rather small-scale event in Bela Crkva became "the day when the first insurgent rifle fired".[4] It acquired enormous importance as the beginning of the organised uprising in Serbia, with the mobilisation speeches by the Partisans and the act of shooting the gendarmes as equally relevant aspects. The assassination fit into the broader war narrative that transcended the actual event in Bela Crkva, interpreting 7 July through the lenses of the entire war, liberation and successful socialist revolution that ensued.

The central celebration took place in Bela Crkva and gathered between 10,000 and 20,000 people, but the Day of the Uprising was celebrated across Serbia. The holiday represented an opportunity for local communities to reveal new memorials, open museums and decorate their local veterans, and municipalities and counties commemorated the beginning of the Partisan uprising in their areas. A rich cultural program was part of the annual celebrations and families and youth used the holiday occasion to enjoy the nature, have picnics and go camping, as many of the commemorative events happened in mountains and parks.

People did not only attend the commemorations, they also helped finance them. As Karge demonstrates, people of Serbia collected 26 million dinars for building local memorials for the tenth anniversary in 1951 and these funds resulted in 212 memorials, 649 plaques and 26 memorial fountains and pyramids (2010: 56). While it is fascinating that people were willing to donate money during the first decade after the Second World War, it was not political conformism or social pressure that motivated them. More often, it was the loss of a family member in the war. Hence, most of the monuments financed through individual donations were expressions of mourning the dead and not future-oriented celebrations of the revolution.

The first celebration of the Day of the Uprising in Bela Crkva took place in 1945 and Tito and the highest Yugoslav, Soviet and other foreign officials attended. In a similar way to other commemorations, the celebration resembled a large village fair with food, music and fireworks. The houses in the village were freshly painted for this occasion. The flags of the western Allies and the Soviet Union were on display around the place, including a banner which said: "Long live great Stalin, the organiser of the victory against fascism!" (*Politika*, 1945). Tito and others celebrated the alliance with the Soviet Union and its people, including praise for Joseph Stalin by the speakers and the audience. Following the general tendencies in politics and memorialisation, the references to the Red Army and the Soviet Union disappeared after 1948. While commemoration speeches always referred to the pressing issues in the economy, internal and foreign affairs, it was always against the backdrop of the grand narrative of Yugoslavia continuing the emancipatory tradition of 7 July.

There are three main aspects of the memory of the beginning of the uprising that reflect the broader characteristics of the Yugoslav memory culture. First, the interpretation of the act of killing two gendarmes by the Partisans was in collective terms. Španac personified the people of Serbia and his action the moment when the Serbian people "screamed loudly that they loved freedom above all" and "took arms and fought their way to the victory", as newspapers reported. In a similar manner, the Memorial to the Uprising in Bela Crkva by Bogdan Bogdanović includes the inscription: "This is where Serbia said freedom" (*Ovde je Srbija rekla sloboda*). With Španac personifying the Serbian people, the shot gendarmes are nameless servants of the occupation, and discourses surrounding the anniversaries never mention their names. The break with the Kingdom of Yugoslavia is another important reference in these collective terms, representing the 1941 uprising as the popular rebellion after 23 years of oppression.

Further, the Yugoslav memory of the 1941 Partisan uprising in Serbia framed it as a continuity with the uprisings and battles in Serbian history, most notably with the 1804 Serbian Uprising against the Ottoman Empire and the 1389 Kosovo Battle. In 1941, the Serbian people "strongly took the rudder of destiny in their hands once again and continued through history following the path they had strived for centuries long" (Petrović, 1945). As Tito emphasised, it was through the People's Liberation War that "the sons of the Serbian nation became worthy of the traditions of their ancestors, who had fought for freedom of Serbia and Serbian people for centuries" (Broz Tito, 1959a: 319).

Finally, the commemorative discourses constantly emphasised that the Communist Party of Yugoslavia planned and prepared the uprising as an inseparable part of the Yugoslav socialist revolution. This also reflects the dominant war narrative based on the unquestionable paradigm of interpretation that represented the Communist Party as the only force consistent in opposing the occupation and collaborationist forces, and as the only option for Yugoslavia and its people (Höpken, 1999a: 223).

Reception of memory politics beyond compliance and rejection

Yugoslav memory culture cannot be reduced to the dichotomy of the imposed Partisan myth and repressed or forbidden memory. The actors of memory politics were not homogeneous either, not even within SUBNOR, especially as the increasing state decentralisation diversified the official memory. Even though the Partisan myth dominated the memory culture, it was not a homogeneous sphere allowing either compliance or rejection and people interacted diversely with the official framework of memory. While some people actively participated in memory work, many were completely indifferent. Partisan monuments and cemeteries were often entirely neglected and sometimes vandalised on purpose. As in every society, vernacular memories thrived and went from mourning the war dead, as opposed to the dominant revolutionary remembrance, to commemorations of collaborators and the war actors deemed people's enemies.

While SUBNOR saw itself as a gatekeeper of memory and of brotherhood and unity, the gate was actually wide open for diverse forms and contents of war memory that run contrary to the officially postulated paradigms (Karge, 2010: 43). Still, people massively and actively participated in the always ongoing creation of war memory promoted by SUBNOR, by attending memorial revealing ceremonies, annual commemorations, school trips and through active preservation of sites of memory (Bergholz, 2006: 82). On the other hand, the Yugoslav memory culture, both as the official and socially practiced war memory, was not a static and monolithic space and diverse appropriations were possible while not leaving the frameworks of dominant memory (Karge, 2010: 23). Instead of simply accepting or rejecting the official "mnemonic offers" (Karge, 2010: 34), people interacted with them in various ways, often transforming and appropriating the dominant narratives and practices. For instance, local practices of remembrance transformed the future-oriented official war narrative of the war, leading to a new and better future to mourning the dead and focusing on loss and pain (Karge, 2009: 51–52).

It has become commonplace to argue that it was only private and emigration circles that cherished the war memories opposed to the official narrative, whispering them in privacy while the state imposed the Partisan myth from above. Then, in the 1980s, the counter-memories suddenly exploded. This interpretation neglects not only the active participation of people in memory work in Yugoslavia, but also the fact that commemorations of the occupiers' domestic helpers existed. Counter-memories existed and were often expressed. The local population often took care of, for instance, Ustasha cemeteries and graves, while neglecting a Partisan cemetery (Bergholz, 2006: 88). Families also gathered to commemorate the members they lost, who were combatants and people's enemies. There are numerous examples of attempts to add names of collaborators to the plaques or memorials to fallen Partisans. In Montenegro, a monument was erected to a Chetnik commander among the Partisans' graves at a village cemetery. One of the examples Bergholz reveals is the local initiative for a memorial plaque supported by a priest in a village near Čačak in Serbia in 1956, dedicated to locals who had fought and died in the Balkan Wars and the First and the Second World War, including 11 Chetnik names. The inscription said: "Those who gave their lives for freedom and unification of the Serb people from 1941 to 1947" (Bergholz, 2010: 14). Someone reported the plaque to the authorities and a discussion and trial followed, sentencing priest Tihomir Veličković to 20 months, after which he returned to the priesthood in Bajina Bašta.

Indifference was more common than active participation or opposition to the official war narrative, as evident in numerous neglected memorials and cemeteries. People neglected memorials, they questioned the dominant narratives and even deliberately destroyed and damaged memorial sites. They also used the sites of memory for diverse purposes: as pastures for their cattle, as ideal places to plant cabbage, and other vegetables, and locations for intimate meetings (Bergholz, 2006).

Remembering the collaboration

The ethnic neutralisation of the Second World War in the dominant memory meant that all Yugoslav nations contributed equally to the antifascist struggle and to collaboration, without any group blamed more than another for inter-Yugoslav atrocities. Glory and guilt were distributed equally to all nations (Höpken, 1999a: 224). Promoting a certain national balance in historical representations meant underlining that all nations had their traitors (Sindbæk, 2009: 51).

The Yugoslav memory entrepreneurs worked on shaping the understanding of the war actors through two binary categories. The first encompassed the men and women who had fought for the creation of the new socialist state as heroes, martyrs and revolutionaries. Those who these heroes, martyrs and revolutionaries fought against were enemies, traitors and counter-revolutionaries (Bergholz, 2010: 36). The reduction of the complex and multi-sided conflict in Yugoslavia implied unconditional resistance and unconditional collaboration (Sundhaussen, 2004: 378). This binary lens placed all political and armed forces who had collaborated with the Axis occupation in one category without differentiation regarding their ideology, the type and level of the collaboration or committed atrocities. Those blamed for the inter-Yugoslav conflicts were the occupation forces, the ruling class and bourgeoisie (Sindbæk, 2012: 75). The domestic "people's enemies" also carried the blame. This group consisted of the Ustasha and Croatian Home Guard (*Domobrani*) in Croatia, the Slovene Home Guard (*Domobranci*), the Chetniks from various parts of the country and the military and police apparatus of the Serbian quisling government of Milan Nedić.

The Chetniks of Dragoljub Mihailović belonged to the category of traitors and collaborators. They were "traitorous and untrustworthy reactionary nationalists", equated with the Ustasha "as collaborators and bestial servants of the Nazi and Fascist occupiers" (Sindbæk, 2009: 51). The Chetniks were both the counter-image to the Partisans and the Serbian counterpoint to the Croatian Ustasha (Sindbæk, 2009: 51). Tito and other state officials usually mentioned the names of Dragoljub Mihailović and Ante Pavelić together, equalising them with other Serbian collaborationist forces when talking about the war in the Serbian context:

> The occupiers realised later, when this struggle began, that the Serbian people could not be seduced to go into the fight against the Croatian people, but they went against the occupier, to fight them as their oppressors. When they realised that they failed, when they saw that the Serbian people took arms in their hands and turned against them, they found Nedić, Ljotić, and later Draža Mihailović too. All of them together, they did one and the same work. Nedić and Draža Mihailović have countless Serbian lives on their conscience, the lives of young men and women, children, and the elderly. Those are not hundreds, not thousands. Those are dozens of thousands.
> (Broz Tito, 1959b: 229)

Equalisation of the Chetniks with other Partisan enemies was, however, not entirely straightforward. Yugoslav authorities emphasised their official position as the representatives of the royal government in exile enjoying the Allied support for most of the war as especially negative, particularly during the regime consolidation in 1945 and the 1946 Belgrade Process against Mihailović. In his 1945 speeches, Tito criticised Mihailović for leading alleged heroes, "those who wanted to get a laurel wreath cheaply, who went to forests and sat there quietly, waiting for the war to end so that they could come out as liberators"

(Broz Tito, 1959c: 62). For Tito and in hegemonic war narratives, the perfidy of the Chetniks did not start at Ravna Gora or anywhere else in Yugoslavia, but in London, where the exiled government sat (Broz Tito, 1959b). In a similar manner, Miloš Minić, the State Prosecutor at the 1946 trial against Mihailović, called him "the most shameful traitor of our people in history" (*Politika*, 1946: 6).

When history outpoured

> Tito's death in May 1980 changed very little for Yugoslavia on the surface. The transition of power to the collective presidency was smooth, the agreed strategy of the leadership was continuity along Tito's path and the lavish state funeral of Yugoslavia's late leader symbolised the commitment of his successors to maintaining the status quo. Appearances hid underlying tensions, however.
>
> (Dragović-Soso, 2002: 64)

The book that opened the floodgates of historical revisionism in Yugoslavia was the 1981 biography of Tito by Vladimir Dedijer, the official chronicler of the Partisan war and Tito's handpicked biographer (Dragović-Soso, 2002: 78).[5] The disclosures in the book opened the doors to challenges to both the Yugoslav revolution and its leader, revealing details of Tito's personal life, his participation in the 1914 Austro-Hungarian attack on Serbia or the role he had played in the arrests of Yugoslav communists in Moscow in the 1930s (Dragović-Soso, 2002: 78). The significance of Dedijer's book goes beyond Tito, however, and it discusses the Second World War and discloses the wartime summary executions and Partisans' negotiations with Germans in March 1943. This "anti-biography" (Ramet, 2005: 322) marks only the beginning of the drastic reinterpretation of the history of the Communist Party of Yugoslavia with the focus on the Second World War period.

In the decade following the death of Josip Broz Tito in 1980, Yugoslavia went through a long political, economic and legitimacy crisis that went hand in hand with the rapidly increasing challenges to the official interpretation of history. The decentralisation of the Yugoslav state had already resulted in increasingly national interpretations of history and historiographies written from perspectives of individual republics and their titular nations, while the Yugoslav dimension had been losing significance. However, the process of nationalising memory in the 1980s, as Höpken explains, fragmented and disintegrated the Yugoslav memory culture, while the same process was slowly happening to the Yugoslav state, its institutions, ideology and political order (1999b: 206). The reversion of the ethnic neutralisation of the war, the Yugoslav efforts of neutralising the negative sides of the common past, gave back the national belonging to victims of fascist terror, as well as the perpetrators (Ristović, 2013: 140).

It was writers, philosophers and jurists, rather than professional historians, who were the first to embrace the topics that challenged the Communist Party's claim to historical legitimacy (Dragović-Soso, 2002: 65; Höpken, 1999b: 205).

In Serbia, the themes that emerged in this period centred on the idea of permanent victimisation of the Serbian nation and "the stab in the back" by communists, interpreting the common Yugoslav state as "the vehicle for the Serbs' destruction as a nation" (Dragović-Soso, 2002: 64–65). In addition to the Serbian nationalist narratives of victimisation, both during the Second World War and in the Yugoslav state, the discussions that gained momentum addressed the repressive state mechanisms of the postwar period and in the years after the split with the Soviet Union.

A wave of literary works appeared already in the early 1980s, dealing with the Serbian victimhood either in the Independent State of Croatia, the postwar repression or the Goli Otok prison camp.[6] In parallel to the Communist Party losing legitimacy, other actors of the Serbian history began to regain significance in society "as memoirs, speeches and writings of the many 'bourgeois' political figures, military leaders and even King Petar himself flooded the market" (Dragović-Soso, 2002: 88). Mass media, as well as historians and other intellectuals, got seriously involved in the discussions on the past and the creation "of a historical discourse in which the borderline between academic historiography and a nonprofessional historical journalism rapidly began to disappear" (Höpken, 1999b: 206).

The Serbian victimisation narrative that became prominent in the public and a subject of controversy focused on the genocide against Serbs in the Independent State of Croatia. In this case, too, it was literary authors who opened the discussion, not historians. The number of victims represented an especially important topic of debate between Serbs and Croats who contested the official numbers by exaggerating and minimising them respectively (Kolstø, 2011). Serbian intellectuals, such as historian Vasilije Krestić, traced the roots for the crimes against Serbs in the NDH back to the 16th and 17th centuries, accusing the entire Croatian nation of being genocidal (Dragović-Soso, 2002: 112). By the end of the decade, genocide surfaced as the central theme in the public, with the Serbian Academy of Sciences and Arts joining this "media orgy" (Dragović-Soso, 2002: 113). This controversy between Serbia and Croatia would further escalate in the 1990s, when the genocide memory would be utilised to justify the war in Croatia as a sort of a "genocide prevention" (Stojanović, 2011: 221).

While the dominant memory in the public went through drastic transformations throughout the 1980s, the official memory of the People's Liberation War promoted by the state actors did not change significantly. Nationalism as a threat and idea that opposed and could endanger everything the Partisans stood for appears within commemorative discourses in the early 1970s and dominated in the 1980s too. Facing the growing challenges to its memory politics and legitimacy in the 1980s, the regime responded by introducing a law protecting Tito's image and legacy in 1984 and repressive actions such as book bans and party expulsions. These measures often had an opposite effect, sparking even greater interest in the writings that the state authorities deemed problematic.

The Second World War reconceived

The decade following Tito's death was the decade of "the outburst of history" (Dragović-Soso, 2002: 77). The narratives that questioned not only the official war memory but also the very existence of the Yugoslav state and its political order thrived in the public sphere, expressed, promoted and discussed like never before. In addition to the narratives of Serbian victimhood and their sacrifice in the Partisan movement, the reinterpretations and revaluations of the nature of the Second World War and its main actors, completely opposed to the dominant narrative of the People's Liberation War, were everywhere. Although its significance and prominence were far beyond the academic context, the paradigm shift was evident in historiography, too.

Published in 1983, Branko Petranović's *Revolution and Counterrevolution in Yugoslavia* was the first historiographic study that shed a slightly different light on the Second World War. The book redefines the Chetniks of Dragoljub Mihailović as the bearers of the resistance against the occupation, referring to them as "antifascist civic forces" (*antifašističke građanske snage*) (Petranović, 1983). He does not ignore the Chetniks' collaboration and critically engages with the émigré and foreign literature that praises the Yugoslav character of the Chetniks and their military strength as the most dangerous German opponents, while depicting the Partisans as a subversive force dependent on Moscow, a foreign centre with the experience of changing the political map of Europe (Petranović, 1983: 11–12). Although the book is a very moderate revision of the dominant paradigm at the time, it is also a novelty because it was the first time in Yugoslav historiography that the term "antifascist" described the Chetniks. Petranović changed his opinion, now arguing that the Chetniks could be conditionally interpreted as an antifascist force before November 1941, their antifascism becoming only declarative later, but he was nevertheless criticised by his colleagues (Biber, 1986; Petranović, 1986a, 1986b).

While Petranović's book did generate critique but did not cause notable public debates, another study that appeared two years later brought the discussion on the nature of the Chetnik movement to the forefront of public attention. This first example of a drastic reinterpretation of the Chetniks in Yugoslav historiography is Veselin Đuretić's *The Allies and the Yugoslav War Drama*, a two-volume study of the Second World War published in 1985 (Đuretić, 1985a, 1985b).

Đuretić explained his work as depicting the drama of national and ideological relations without the usual political regimentation (Đuretić, 1985a: 10). He adopts Serbian nationalism as his lens of war interpretation. In his books, foreign factors and conspiracy are decisive for the war's outcome in Yugoslavia, the Chetniks are antifascists who had to collaborate, and the Partisans and their ideology are foreign to the Serbian people. Antifascism becomes the main characteristic of Serbian people, whether among the Chetniks or the Partisans. Rooted in anti-communism and representing the Chetniks in an exclusively positive light, the arguments Đuretić had outlined became foundations for later generations of historians. The significance of Đuretić's work is providing the basic plot story that became a commonplace in academic and non-academic discourses of Serbian anti-communism and nationalism. His theses will only be further elaborated, reproduced and built upon in the years to come (Bešlin, 2013: 90).

For Đuretić, foreign factors are crucial. According to him, it was only because of the support of the Soviet Union and the western Allies that the Partisans managed to achieve victory (Dragović-Soso, 2002: 102). The Partisans are foreign to Serbia and the Communist Party is international and entirely inadequate to the Serbian context, working on the fulfilment of their main revolutionary goal – the seizure of power. The Allies provided them with the opportunity to fulfil their goals only because of their misunderstanding of the situation (Đuretić, 1985a: 28). For Đuretić, international actors had an anti-Serb attitude that arose from their belief in the greater Serbian hegemony in interwar Yugoslavia, another myth Đuretić wanted to debunk. Đuretić's theory of the foreign conspiracy against the Serbian nation will always gain more prominence and culminate during the wars of the 1990s. The international conspiracy of foreign actors against Serbs in the Second World War context resulted in the betrayal of the Chetniks by the Allies and their turn to the communist Partisans.

For the Serbian nation as inherently antifascist, the Partisans and Chetniks represent "the two military-political expressions of people's spirit" (Đuretić, 1985a: 45). Speaking of the two movements as Serbian antifascist forces, Đuretić explains that his focus on the Serbian positions is not because they were especially patriotic in comparison to other Yugoslav nations, but because Serbs "were targeted by all occupiers as a nation so its antifascism had to be above all ideologies and politics" (1985a: 6). While the Chetniks are consistent antifascists, however, the Partisans consider the war as a framework for revolution. The Chetniks had to collaborate to protect the Serbian people from repression and that was the main motif for "anti-communist compromises" (Đuretić, 1985a: 176). The Partisans, on the other hand, blinded by their revolutionary goals, did not care about retaliations and suffering. In the following decades, Đuretić's idea and the exact phrase of "protection of Serbian people from repression" (*zaštita srpskog naroda od represalija*) would become the most frequently repeated argument in public discourses and revisionist historiography.

Historians such as Đorđe Stanković and Branko Petranović criticised all aspects of Đuretić's war interpretation, accusing him of attempting to deconstruct a myth in order to construct a new one and making internal forces invisible as independent historical subjects by exaggerating the Allies' importance (Stanković, 1985a: 36, 1985b: 40). Criticising Đuretić for distorting projections of historical reality and changing the meaning of already known facts, Petranović justly argued that Đuretić's work was not a historiographic result and it testified more about the present than about the past (1986a: 12). In a failed attempt to defend his work, Đuretić considered himself misunderstood and he rejected accusations for rehabilitation of the Chetniks, arguing that he represented them in all lights (*Politika*, 1986b: 11).

The reaction to the book, rather than the work itself, made *The Allies and the Yugoslav War Drama* widely known, even though its launch at the Serbian Academy of Sciences and Arts involved around 500 very prominent guests. Media, SUBNOR and historians publicly condemned the book. In November 1985, the League of Communists expelled Đuretić for "national intolerance and chauvinism" (*Politika*, 1985). The controversial book generated public discussions, roundtables and feuilletons in 1985 and 1986. The court banned the third edition of the book in December 1985 but the Supreme Court of Serbia abolished the ban. These debates and the ban contributed to even greater public interest in the book and the first two editions had already sold out before the court ban. Because of attacks on Đuretić, his expulsion from the party and the court ban, the Serbian nationalist intellectual opposition mobilised to defend his right to free speech. In this way, he received legitimacy and a scholarly renown he may not otherwise have achieved with his work as a historian (Dragović-Soso, 2002: 103).

For this book, Veselin Đuretić did not work with any important archival sources that professional historians had not considered before. The interpretations his book offers are, thus, not based on the discovery of new sources or on a novel theoretical and methodological approach. In the same manner, the book does not reflect a new paradigm in academic historiography, but is an illustration of the political context of the 1980s in Serbia and Yugoslavia. The book, as Petranović put it, mirrored the time it was written in, rather than revealing something new about the time it examines. Even though *The Allies and the Yugoslav War Drama* does not represent a valuable scholarly contribution to the history of the Second World War, its social and political significance is immense and it outlived the post-Tito decade.

The interpretation of the Second World War and its actors, as outlined in Đuretić's work, embodies the discourses shared by many Serbian intellectuals of the 1980s. During the 1990s, these narratives were omnipresent and Đuretić himself became politically active. More importantly, Đuretić's arguments became the backbone of Serbian revisionist historiography of the late 1990s and 2000s, written by a small group of historians acting as important state-affiliated mnemonic agents. They did not move much further than this 1980s example of history writing in their work. The narrative of Serbian antifascism, involving both Chetniks and Partisans and ascribing decisive importance to the role of the international actors, is the focal point of state-sanctioned memory politics in post-Milošević Serbia. The apparent equalisation of the two movements stands for the image of the Partisans as foreign to Serbia and its people and completely careless about their suffering. The Chetniks, on the other hand, are the movement deeply rooted in the Serbian countryside who enjoyed a wide popular support, caring about the protection of people from repression and reprisals, even if it pushes them into collaboration with their enemies. It was the 1980s when the notion of the civil war within the Serbian nation also became salient in the interpretation of the Second World War in Yugoslavia. The view portraying the Serbian people as antifascist – regardless of whether in the Chetniks or the

Partisans – and as victims of the genocide in the NDH – emphasising the international conspiracy of great powers and of communism against the Serbs that dominated the 1980s debates – spilled over to the tumultuous 1990s.

Notes

1 The last Second World War trial in Yugoslavia took place in 1986. It was against Andrija Artuković, the interior and justice minister and state secretary of the Independent State of Croatia who was responsible for racial laws. He fled Yugoslavia and spent most of his life in the United States. The court sentenced Artuković to death at the trial that attracted enormous public attention, but he was not executed due to his age and health issues. He died in custody in 1988.
2 Odbor za obeležavanje i uređivanje istorijskih mesta iz Narodnooslobodilačkog rata.
3 An alternative date of 22 June was given for the uprising in Croatia that existed during the Yugoslav period and became the Day of the Antifascist Struggle in 1991, replacing 27 July. In 1941, a group of communists from Sisak formed the first Partisan unit in Croatia and Yugoslavia, having heard of the German attack on the Soviet Union. In Yugoslavia, the date was usually referred to in the international context of the attack on the Soviet Union, but it was also locally commemorated in Sisak. It increasingly gained more prominence until it became the central celebration of the Croatian resistance to fascism.
4 Španac used a Steier gun he had in his coat pocket and not a rifle but this phrase recurs in Yugoslav newspapers and speeches of Yugoslav officials already from 1945.
5 Dedijer, V. (1981) *Novi prilozi za biografiju Josipa Broza Tita*. Rijeka: Liburnija.
6 Drašković, V. (1983) *Nož*. Belgrade: Zapis; Isaković, A. (1982) *Tren 2*. Belgrade: Prosveta; Popović, D. (1986) *Knjiga o Milutinu*. Belgrade: Književne novine; Radulović, J. (1983) *Golubnjača*. Belgrade: BIGZ.

Bibliography

Baker, C. (2015) *The Yugoslav Wars of the 1990s*. Studies in European History. Basingstoke: Palgrave Macmillan.
Bergholz, M. (2006) "Među rodoljubima, kupusom, svinjama i varvarima: Spomenici i grobovi NOR-a 1947–1965. Godine". In Kamberović, H. (ed.), *60 godina od završetka Drugog svetskog rata – Kako se sjećati 1945. godine*. Sarajevo: Institut za istoriju, pp. 75–101.
Bergholz, M. (2010) "When All Could No Longer Be Equal in Death: A Local Community's Struggle to Remember Its Fallen Soldiers in the Shadow of Serbia's Civil War, 1955–1956", *The Carl Beck Papers in Russian and East European Studies* (2008): 1–58.
Bešlin, M. (2013) "Četnički pokret Draže Mihailovića – najfrekventniji objekat istorijskog revizionizma u Srbiji". In Samardžić, M., Milošević, S. and Bešlin, M. (eds), *Politička upotreba prošlosti: O istorijskom revizionizmu na postjugoslovenskom prostoru*. Novi Sad: AKO, pp. 83–143.
Biber, D. (1986) "Još jednom o 'antifašizmu' četnika". *Komunist*, 24 January.
Broz Tito, J. (1959a) "Govor na Kosmaju (Povodom dana ustanka naroda Srbije)". In Stanojević, T. (ed.), *Govori i članci*. Zagreb: Naprijed, pp. 319–330.
Broz Tito, J. (1959b) "Govor u Užicu na Dan ustanka u Srbiji". In Stanojević, T. (ed.), *Govori i članci*. Zagreb: Naprijed, pp. 225–234.
Broz Tito, J. (1959c) "Jugoslovenska armija i bivši oficiri-zarobljenici". In Stanojević, T. (ed.), *Govori i članci*. Zagreb: Naprijed.

Dragović-Soso, J. (2002) *Saviours of the Nation? Serbia's Intellectual Opposition and the Revival of Nationalism*. London: C. Hurst & Co. Publishers.

Đuretić, V. (1985a) *Saveznici i jugoslovenska ratna drama: Između nacionalnih i ideoloških izazova*. Belgrade: Narodna knjiga, Balkanološki institut SANU.

Đuretić, V. (1985b) *Saveznici i jugoslovenska ratna drama: Prestrojavanja u znaku kompromisa*. Belgrade: Narodna knjiga, Balkanološki institut SANU.

Höpken, W. (1999a) "Vergangenheitspolitik im sozialistischen Vielvölkerstaat: Jugoslawien 1944–1991". In Bock, P. and Wolfrum, E. (eds), *Umkäpfte Vergangenheit. Geschichtsbilder, Erinnerung Und Vergangenheitspolitik Im Internationalen Vergleich*. Göttingen: Vandenhoeck & Ruprecht, pp. 210–247.

Höpken, W. (1999b) "War, Memory, and Education in a Fragmented Society: The Case of Yugoslavia", *East European Politics and Societies* 13(1): 190–227.

Höpken, W. (2006) "Jasenovac – Bleiburg – Kočevski Rog: Erinnerungsorte als Identitätssymbole in (Post-)Jugoslavien". In Richter, A. and Beyer, B. (eds), *Geschichte (ge-)brauchen. Literatur und Geschichtskultur im Staatssozialismus: Jugoslavien und Bulgarien*. Berlin: Frank & Timme, pp. 401–433.

Horvatinčić, S. (2011) "Formalna heterogenost spomeničke skulpture i strategije sjećanja u socijalističkoj Jugoslaviji", *Anali galerije Antuna Augustinčića* (31): 81–106.

Judt, T. (2005) *Postwar: A History of Europe Since 1945*. New York: The Penguin Press.

Karge, H. (2009) "Mediated Remembrance: Local Practices of Remembering the Second World War in Tito's Yugoslavia", *European Review of History: Revue européenne d'histoire* 16(1): 49–62. DOI: org/10.1080/13507480802655394.

Karge, H. (2010) *Steinerne Erinnerung – versteinerte Erinnerung? Kriegsgedenken in Jugoslawien (1947–1970)*. Wiesbaden: Harrassowitz.

Kirn, G. (2016) "Transnationalism in Reverse: From Yugoslav to Post-Yugoslav Memorial Sites". In de Cesari, C. and Rigney, A. (eds), *Transnational Memory. Circulation, Articulation, Scales*. Media and Cultural Memory 19. Berlin: De Gruyter, pp. 313–338.

Kolstø, P. (2011) "The Serbian–Croatian Controversy over Jasenovac". In Ramet, S. P. and Listhaug, O. (eds), *Serbia and the Serbs in World War Two*. Basingstoke, New York: Palgrave Macmillan, pp. 225–246.

Koren, S. (2012) *Politika Povijesti u Jugoslaviji (1945–1960): Komunistička Partija Jugoslavije, Nastava Povijesti, Historiografija*. Zagreb: Srednja Europa.

Manojlović Pintar, O. (2010) "(Ne)vidljiva mesta sećanja. Spomenici crveenoarmejcima u Srbiji". In *Oslobođenje Beograda 1944*. Belgrade: Institut za noviju istoriju Srbije, pp. 545–553.

Milošević, S. (2017) "Yugoslav Society 1918–1991: From Stagnation to the Revolution". In *Yugoslavia from a Historical Perspective*. Belgrade: Helsinki Committee for Human Rights in Serbia, pp. 349–391.

Petranović, B. (1983) *Revolucija i kontrarevolucija u Jugoslaviji (1941–1945)*. Belgrade: Izdavačka radna organizacija Rad.

Petranović, B. (1986a) "Istoriografija 1: Fetišizam izvora i stvarnost". *Komunist*, 10 January.

Petranović, B. (1986b) "Istoriografija 2: Fetišizam izvora i stvarnost". *Komunist*, 17 January.

Petrović, M. S. (1945) "Sedmi jul – dan ustanka u Srbiji". *Politika*, 7 July. Belgrade.

Politika (1945) "Veličanstvene svečanosti u Beloj Crkvi, na Kosmaju i u Beogradu". 7 August. Belgrade.

Politika (1946) "Dvadeset šesti dan suđenja Draži Mihailoviću i ostalim izdajnicima". 9 July. Belgrade.

Politika (1985) "Dr. Đuretić izbačen iz Saveza komunista". 6 November. Belgrade.

Politika (1986a) "Ideološke projekcije i stvarnost". 11 January. Belgrade.
Politika (1986b) "Nedopustive simetrije". 16 January. Belgrade.
Ramet, S. P. (2005) *The Three Yugoslavias: State-Building and Legitimation, 1918–2005*. Bloomington: Indiana University Press.
Ristović, M. (2013) "Kome pripada istorija Jugoslavije?" *Godišnjak za društvenu istoriju* XX(1): 133–143.
Roksandić, D. (1995) "Shifting References: Celebrations of Uprisings in Croatia, 1945–1991", *East European Politics and Societies* 9(2): 256–271. DOI: org/10.1177/0888325495009002004.
Sindbæk, T. (2009) "The Fall and Rise of a National Hero: Interpretations of Draža Mihailović and the Chetniks in Yugoslavia and Serbia since 1945", *Journal of Contemporary European Studies* 17(1): 47–59. DOI: org/10.1080/14782800902844693.
Sindbæk, T. (2012) *Usable History? Representations of Yugoslavia's Difficult Past from 1945 to 2002*. Aarhus: Aarhus University Press.
Stanković, Đ. (1985a) "Jugoslavija bez simetrije 1". *NIN*, 22 September.
Stanković, Đ. (1985b) "Jugoslavija bez simetrije 2". *NIN*, 29 September.
Stojanović, D. (2011) "Value Changes in the Interpretations of History in Serbia". In Ramet, S. P., Listhaug, O. and Dulić, D. (eds), *Civic and Uncivil Values: Serbia in the Post-Milošević Era*. Budapest, New York: Central European University Press, pp. 221–241.
Sundhaussen, H. (2004) "Jugoslawien und seine Nachfolgestaaten. Konstruktion, Dekonstruktion und Neukonstruktion von 'Erinnerungen' und 'Mythen'". In Flacke, M. (ed.), *Mythen der Nationen. 1945 – Arena der Erinnerungen*. Berlin: Deutsches Historisches Museum, pp. 373–427.
Velikonja, M. (2017) "Ways of Remembering Yugoslavia. The Yugoslav Rear-View Window". In *Yugoslavia from a Historical Perspective*. Belgrade: Helsinki Committee for Human Rights in Serbia, pp. 515–551.

4 The Milošević era

> Milošević's simultaneous appeal to very different constituencies (not just nationalism) was key to his political success: Yugoslavia, unity, and Titoism for the party orthodox and army officers; Serbia for the nationalists and state support for the Kosovo Serbs; reform and rehabilitation for the intellectuals, social justice and protection for state-dependent workers and pensioners.
>
> (Vujačić, 2004: 30)

In 1997, Zoran Đinđić of the Democratic Party (*Demokratska stranka*, DS) became the mayor of Belgrade. The state did not acknowledge the local election results at first, causing massive protests in Belgrade and across Serbia. After the intervention of the Organisation for Security and Cooperation in Europe (OSCE), Slobodan Milošević had to accept the electoral victory of the opposition, gathered in the *Zajedno* coalition. Belgrade had its "first democratically elected mayor" (*Blic*, 2017). The coalition would disintegrate and Đinđić would be ousted from the office by autumn, but, in February 1997, there were many reasons for celebration. Having protested for over three months in the cold winter, the coalition and citizens who opposed the Milošević regime got together to celebrate the constitution of the Assembly of the City of Belgrade. The festivities culminated in the removal of the concrete five-point star from the dome of the Assembly building that had been there since the postwar period. The alpinists, hired in advance for this occasion, took the star down and handed it over to the newly elected major. The photograph of Zoran Đinđić holding the fallen five-point star, the symbol of communism, became one of the most legendary images of 1990s politics in Serbia.

Ten years before that night, Slobodan Milošević took control of the Serbian branch of the League of Communists, marking the beginning of the era that lasted until 5 October 2000.[1] Under the leadership of Slobodan Milošević in the late 1980s and early 1990s, Serbia went through the transition from communism to "national authoritarianism" from above (Ostojić, 2014: 23). The regime that replaced Yugoslav state socialism in rump Yugoslavia was a hybrid regime that employed authoritarian practices behind the façade of democracy, involving pluralist elections and "proto-democratic institutions" but without any real alteration of power (Ostojić, 2014: 21). It was not a straightforward totalitarian dictatorship, but "a scrupulous subversion of democracy orchestrated by segments of the old communist elites that endorsed nationalism" (Ostojić, 2014: 22). It was a "patronage-based and criminalised form of capitalism" (Mikuš, 2018: 53).

The first elections at the level of the republic took place in 1990. The Socialist Party of Serbia (*Socijalistička partija Srbije*, SPS), as the League of Communists of Serbia renamed themselves, enjoyed relatively broad support and Milošević offered the status quo, but they also had numerous institutional advantages and control over electronic media (Uvalić, 2010: 42). The opposition parties that started forming before the elections stood no chance against it. In addition to the personal popularity of Milošević, the SPS dominated both the federal and the Serbian parliament and had full control of media, the local government organisations and the police (Uvalić, 2010: 46).

The regime's authoritarian nature became obvious very quickly. The SPS showed its willingness to suppress any challenges to its rule by force already in March 1991 when the regime crushed a mass opposition protest calling for liberalisation of the media, sending army tanks into the centre of Belgrade. This resulted in the shutting down of the independent and student television and radio station and the arrest of Vuk Drašković of the Serbian Renewal Movement (*Srpski pokret obnove*, SPO) (Uvalić, 2010: 42).

The SPS regime laid stress on continuity with Yugoslav state socialism, although it had formally moved away from it already in its early days. The shift from state socialism occurred in 1989 with the erasure of the workers' self-management system and the legalisation of privatisation of public property, with the goal of adjusting to the international context and enabling foreign investments (Woodward, 1995: 280). The constitution of 1990, that remained in effect until 2006, formalised this shift by removing "socialist" from the name of the state and "instituting the liberty and legal equality of private ownership and business, though without scrapping 'social' property" (Mikuš, 2018: 56).

The memory culture involved the SPS regime representing itself as the direct successors of socialist Yugoslavia and keepers of the memory and tradition of the People's Liberation War. They, however, "ethnified" the Partisan myth as Serbian, deprived the Partisan struggle of its revolutionary dimension and narrated it through the dichotomy of Serbian parallel victimhood and heroism. In the context of the wars of the Yugoslav disintegration, the regime used the discourses of the People's Liberation War to position themselves as a bulwark against fascism, fascism being the secessionist struggle of non-Serbian Yugoslav nations against the dissolving Yugoslav state. At the same time, the regime tolerated the Chetnik revival that came from the opposition parties and gained momentum in the wars in Croatia and Bosnia.

Changes and continuities

The memory politics of the Milošević's regime was as ambiguous as its ideology. As the multiparty system formed before the 1990 elections, the SPS leadership "had to balance ethnonationalist traditionalism and socialism in order to make ideological gain against much more nationalist opposition parties" (Malešević, 2002: 178). The narratives that the Milošević regime promoted were ethnic (Serb) and occasionally religious (Serbian Orthodox) (Malešević, 2002: 177). The historical debates of the 1980s spilled over to the 1990s and gained even more prominence, particularly the Serbian victimhood in the NDH and the negative position of Serbs in socialist Yugoslavia. Moreover, the political relevance of the Kosovo myth intensified at the official level after the Gazimestan speech by Slobodan Milošević on the 600th anniversary of the Kosovo battle in June 1989 (Bieber, 2002). The stories of Serb victimisation and heroism dominated the public sphere. During the wars in the 1990s, state actors utilised the national myths, heroes and symbols to both justify and mobilise for the wars.

The paradox of the SPS regime was that it employed nationalist propaganda and endorsed wars, while simultaneously insisting on representing the continuity with Yugoslav socialism and promoting the idea of socialism and democracy as inseparably intertwined political values (Malešević, 2002: 187). This reflected in memory politics too. The SPS continued with the Yugoslav practices of celebrating the notable dates of the People's Liberation War, appropriating them for legitimisation of their rule and adjusting it to their ambiguous rhetoric. The aim of the regime was, in fact, to rescue Serbian identity from the perceived threat of the Yugoslav one (Subotić, 2015: 190). In the commemorative context, this meant that other Yugoslav people who had fought in the Partisans became invisible. Instead of them, the delegations of Belarus, Russia and Ukraine attended commemorations of the Victory Day or the Day of the Liberation of Belgrade in the 1990s, as the nations who had contributed to the victory against fascism together with Serbs and the more suitable allies (*Borba*, 1998a: 2).

The commemorations of the People's Liberation War helped maintain the narrative of Milošević's regime as the successor of socialist Yugoslavia and the Partisans' struggle. While the SPS succeeded the League of Communists of Serbia, the Serbian branch of SUBNOR continued existing in its role as a veteran association and a mnemonic agent, now with a predominantly ceremonial role in commemorations, but supporting the regime. The associations of veterans of the 1990s wars would later become members of SUBNOR, arising from SUBNOR's conviction, shared with the SPS regime, that rump Yugoslavia represented the defender of the freedom in the context of the wars and imposed sanctions. The regime used the narrative of continuity with the People's Liberation War to represent themselves as fighting for the freedom of (Serbian) people, such as in the official SPS felicitations for the Victory Day in 1998:

> The People's Liberation War and defending the country, the resistance to fascism, and enormous sacrifices and destruction we had because of it, placed our country to a high position among the countries of the antifascist coalition, because our people have never in its history had doubts when their freedom, independence, honour, and dignity had to be defended. Today as well, Serbia and Yugoslavia stand firm in defending their freedom, independence, and territorial integrity, in their orientation towards

peace and cooperation on equal basis and for the further economic, cultural, and universal social development.

(*Borba*, 1998c)

In the circumstances of state disintegration and the wars in the first half of the 1990s, the state actors used the People's Liberation War as a metaphor to depict all forces struggling for succession from the common Yugoslav state as fascist, while Milošević's Yugoslavia stood as a bulwark against the revival of fascism.[2] The Yugoslav Army, acting as the main successor of the People's Liberation Army, played a crucial role in all commemorations of the Second World War. These commemorations often went hand in hand with military parades and air shows. In this manner, General Momčilo Perišić as the chief of army congratulated the people of Serbia on Victory Day in 1998, explaining that the Yugoslav Army celebrated 9 May as its own holiday,[3] "continuing the tradition of the victorious army of the two Balkan and world wars" (*Borba*, 1998b: 3).

In addition to the ethnicisation of the People's Liberation War and its actors, the main Partisan commander, Josip Broz Tito, became unsuitable for the dominant memory promoted by the SPS. On the anniversary of Tito's death in 1992, the Presidency of Yugoslavia decided to remove the honour guard from his mausoleum and replace it with the security staff of the memorial centre. This was the first year that the House of Flowers stayed closed on the anniversary and no official delegations visited it to lay wreaths, as they had done every year previously (*Politika*, 1992: 11).[4]

The state-sanctioned history textbooks echoed the dominant ideology and hegemonic discourses, "using the same socialist matrix and language, with super-imposed ethno-nationalist layers" (Pavasović Trošt, 2018: 10). Reflecting the commemorative discourses, Tito disappeared from textbooks and became a nameless figure of "the military commander". The 1990s textbooks highlight the great contribution of the Serbian people to the victory against fascism as well as their suffering (Pavasović Trošt, 2018: 13).

The anti-communist opposition

While the official memory politics combined ethnonationalism and the Partisan myth, commemorations of the defeated side of the Second World War, namely the Chetnik movement, thrived within a large segment of the political opposition. The opposition parties emerged before the 1990s elections, many of which would come to power in 2000, after a decade of being in opposition and in different compromises with the SPS. With notable differences in the political orientation and programs, almost all opposition parties were staunchly anticommunist. Their manifestos prioritised condemnation of Yugoslav state socialism and plans for overcoming its legacies. A liberal opposition to Milošević in the form of civil society existed as well. While they criticised the violations of democracy and nationalist and populist rhetoric of the regime and were often connected to the opposition (Mikuš, 2018: 57), they did not engage in anti-communist memory work or investigation of communist crimes, such as the postwar executions or post-1948 persecution of alleged pro-Soviet elements.

The first manifestos of political parties foregrounded their anti-communist position and promoted the narrative of national reconciliation. One of them was the 1990 program of the Serbian National Renewal (*Srpska narodna obnova*), the predecessor of the SPO led by Vuk Drašković. The party had a goal of ending the civil war among the Serbian people that started on 7 July 1941 (Drašković, 1990: 22). The program argued for the renewal of all Serbian people's values that had been "violently and systematically exterminated, dissipated and supressed" from 1941 on, first by the Ustasha, the occupiers and the civil war, and later by "the Bolshevik-Titoist dictatorship of a few and their strategy of crippling the Serbdom" (Drašković, 1990: 20). In a similar way to the Chetnik wartime plans, the party advocated Greater Serbia within Yugoslavia and ethnic homogenisation. A special section of the program dealt with Tito as the leader of the anti-Serbian strategy. Similarly, the Serbian Freedom Movement (*Srpski slobodarski pokret*) of Vojislav Šešelj proposed the removal of "the pagan temple officially called the House of Flowers" and transferring Tito's remains to his birthplace in Croatia, where he belonged (Otadžbinska centralna uprava srpskog slobodarskog pokreta, 1990: 28).[5]

Although their statement of intent seems moderate in comparison to other parties that were more radically nationalist, the DS also stressed the negative role of Tito as "one of the biggest profligates among the rulers of the modern world" (Demokratska stranka, 1990: 24). The reestablishment of the DS in 1990 was deeply rooted in the narrative of communism as the worst thing that had happened to Serbia and its people (Milošević, 1990; Pekić, 1990). In the same way as the future SPO, the DS would fight for national reconciliation as "the only path towards the end of the civil war that our people went through during the Second World War" (Stojanović, 1990). A legal scholar and DS member at the time, Kosta Čavoški, who acted as one of the plaintiffs in the rehabilitation process of Dragoljub Mihailović in the post-Milošević period, summarised the interpretation of Yugoslav socialism as the long, dark night of national history that the DS stood for:

> The human spirit has its deep nights that ruin the work of its days. That night when our freedom was eradicated, and the most modest form of European parliamentarism and responsible government was destroyed, was the postwar communist totalitarianism.
>
> (Čavoški, 1990)

The SPO led by Vuk Drašković had anti-communism as a defining characteristic, in addition to the monarchism that separated this party from the rest of the opposition. They also promoted national reconciliation through ending the civil war and abolishing the division between the Chetniks and the Partisans (Ćirić, 1990: 31). The SPO became the most vocal opposition party to Slobodan Milošević that attracted massive support. Furthermore, they also represented the most powerful mnemonic agent centred on the celebration of the Chetniks and Dragoljub Mihailović.

The media also promoted revision of the past. The most notable example was *Pogledi* with Miloslav Samardžić as the editor-in-chief, but the weekly *NIN* also often published articles about the Chetniks (Đuretić, 1990). Originally a monthly student newspaper of the University of Kragujevac established in the early 1980s, *Pogledi* grew into a prominent political magazine. It had numerous prominent contributors such as Biljana Srbljanović, Veran Matić and Vanja Bulić, and its sales went up to 200,000 in the early 1990s. With Samardžić being a hobby historian, the historical topics prevailed as *Pogledi* became increasingly and more openly political.

In the late 1980s and early 1990s, *Pogledi* published several feuilletons about the Chetniks, with every issue covering postwar retribution in a city or region of Serbia. The magazine collaborated increasingly closely with Serbian political emigration, particularly the Chetniks. In the early 1990s, Pogledi started openly propagating the reestablishment of monarchy, the return of the Karađorđević family and the reconstruction of the interwar monuments of King Aleksandar that had been destroyed in postwar Yugoslavia (Samardžić, 1990). As a publishing house that still exists, *Pogledi* republished Chetnik memoirs previously published abroad and dozens of books about the Chetniks written by Miloslav Samardžić. Taking the struggle against communism even further, Samardžić and *Pogledi* initiated a series of war crimes lawsuits against former Partisans, which Serbian courts rejected.[6]

Between the Chetnik revival and commemorations of the postwar retribution

While the SPS regime continued the commemorative practices of socialist Yugoslavia and represented themselves as defenders of antifascist values, instrumentalising the Yugoslav memory culture in the context of state dissolution and armed conflicts, the political opposition commemorated the defeated forces of the Second World War and the groups they considered victims of communism. The state authorities tolerated the growing popularity of the Chetnik movement that was not only limited to a few political parties, but resonated wider in society. Markets and street stalls across cities in Serbia were swarming with all sorts of Chetnik memorabilia, the Chetniks' songs were recorded and sold on cassette tapes and Ravna Gora mountain attracted thousands of people every May to commemorate the 1941 establishment of the Chetnik movement. The revival of the Chetnik movement in the wars included Serbian combatants and paramilitaries in Croatia and Bosnia wearing Chetniks imagery, often resembling the Chetniks from the Partisan films. State-sanctioned textbooks revised the Second World War moderately, representing the Chetniks as both antifascists and quislings (Pavasović Trošt, 2018: 13).

In a similar way to the SPS claiming the continuity with the People's Liberation War, the SPO claimed the Chetniks and appropriated their traditions. For the

SPO, "the Serbian resistance movement that had begun at Ravna Gora in 1941 continued in 1990 as the Serbian Renewal Movement" (Stefanović, 1992: 78). After the 1990 split, Šešelj's Serbian Radical Party accused the SPO of appropriating the Chetniks for their political promotion and they did not participate in any SPO-affiliated commemorative practices. When the SPO erected the statue of Dragoljub Mihailović at Ravna Gora in 1992, Aleksandar Stefanović of the SRS attacked them for stealing his idea, arguing that the monument would represent a new division of the Serbian nation because it was not the Serbian people who erected the monument, but one political party whose leader wanted popularity (Stojić, 1992: 8).

The DS did not officially commemorate the Chetniks or other defeated Second World War forces as a party. Its prominent members, however, took part in commemorations. At one of the first commemorations of victims of communism held in Vavedenjska Church in Belgrade in May 1990, Vojislav Koštunica, a founding member of the new DS, held a speech (Stefanović, 1992). He would become the president of Yugoslavia and prime minister of Serbia in the post-2000 period.

Religious memorial services, such as the one Koštunica attended, became a vital part of commemorations of victims of communism that surfaced in 1990, often involving the highest clergy of the Serbian Orthodox Church. The Church was involved in the anti-communist work from the beginning and religious memorial service remains the usual commemorative practice of this community of memory in Serbia today, too. The clergy takes part in commemorations and performs the memorial service and the Church often allows the building of memorials to victims of communism and Chetniks at its property. Inherently anti-communist, the Serbian Orthodox Church emphasises its victimhood under communism in the broader context of the decades of victimisation of the Serbian nation. The religious commemorative practices of the early 1990s were dedicated to the Karađorđević rulers as well. The bishop of Šumadija held the first memorial service to King Aleksandar I since the Second World War in Kragujevac in autumn 1990, that *Pogledi* widely promoted (Bergam, 1990).

State authorities prevented the first attempts of commemoration of the postwar executions or the Chetniks, employing police and army forces. The regime perceived these practices as a threat only before the first multiparty elections in 1990 and targeted large gatherings organised by the newly formed political parties. As the SPS grip on power consolidated, the regime did not approve but it tolerated these practices. In May 1990, however, the state authorities disbanded the first commemoration of the establishment of the Chetnik movement at Ravna Gora, as Vuk Drašković recalls:

> It was not a gathering. It was a hike to a forbidden mountain. After the victory of the communist revolution in Yugoslavia, it was not only the movement of General Mihailović, the Yugoslav Army in the Homeland, and himself as its commander, that were declared criminals. The mountain of Ravna Gora was also declared a criminal, the mountain where the

wartime guerrilla headquarters were. A hundred of us started hiking to Ravna Gora but we could not reach the mountain. As we had said publicly that we would go there, the soldiers of the Yugoslav Army garrison in Valjevo were mobilised and they, masked and with full military equipment, occupied the roads to Ravna Gora. At the same time, strong police forces were also sent, led personally by Minister of Police Radomir Bogdanović. They blocked us, so we could not reach Ravna Gora and erect the prepared plaque that said "In Topola Karađorđe, in Takovo Miloš, at Ravna Gora Draža". We managed to bring the plaque to Ravna Gora a year later.

(Drašković, 2016)

According to media reports, the police knew about the planned gathering and were there to dissolve it only because the SPO had not registered it with the authorities, while Drašković attempted to convince them it was only an ordinary mountain hike (Drašković, 2016). Media reports portrayed the SPO endeavour negatively. For instance, *Politika* called it "a hike into the dark past" and claimed that the initiative had no popular support whatsoever because "everyone still remembered the Chetnik dagger" (Mandić, 1990).

Later that year, the local SPO chapter in Čačak organised a religious memorial service for victims of fascist and communist terror together. The authorities banned the event but the organisers decided to proceed. The police blocked the streets leading to the Borac stadium where the commemoration was about to take place, but between 2,000 and 3,000 people managed to gather anyway. Finally, the memorial service took place, only with a one-hour delay (Radosavljević, 1990). From 1991 on, as the SPO started organising the always-growing gatherings at Ravna Gora, the state authorities did nothing to prevent it and the media did not report on them extensively, but rather ignored them. If pro-regime media reported about Ravna Gora at all, it was in a very negative light.

Ravna Gora gatherings

The 1990 attempt to hike to Ravna Gora by the SPO and its adherents grew into the annual Ravna Gora gathering (*Ravnogorski sabor*) that commemorates the arrival of Dragoljub Mihailović to the mountain and subsequent establishment of the Chetnik movement. Bringing together the individual, small-group and social levels of memory work and diverse memory choreographers, Ravna Gora became the central gathering of the Chetnik memory community in the early 1990s. After the 1990 fiasco of putting up a plaque, the commemoration always grew bigger and the SPO with the support of the Serbian Orthodox Church still organises it today. No other pro-Chetnik actors, such as the SRS, have managed to gather so many people at a commemoration. Even though Ravna Gora gathering gained enormous popular support in Serbia, with thousands attending, the pro-regime media represented it as merely an SPO matter and referred to people who attended as the SPO followers, arguing that the Chetnik celebration did not have wider support in society. As a 1992 newspaper report remarked, the Serbian public "was shamefully silent as if all of this happened on a different planet, as if these were not their people" (Ćirović, 1992: 14).

In 1992, at the ceremony to reveal the statue of Dragoljub Mihailović, thousands of people gathered with "cockades, šajkača hats, flags and miscellaneous fire arms" (Radetić, 1992). The SPO organised 20 buses only from Belgrade. The large volume of firearms at the gathering and their usage in the form of constant firing of guns into the air became a Ravna Gora tradition, inseparable from the commemoration. As one report explained, the police did not dare to prevent the gathering, not because of some ideological reasons, but out of pure fear that the armed volunteers would riddle them "in the spirit of Ravna Gora" (Stefanović, 1992: 78). The SPO saw it differently:

> The surprising passivity of those who were breaking the memorial plaque just two years ago has a possible explanation. We could look for it in the aim of our regime to hear the wishes of its people, but this explanation is far stretched. It is more probable that our dying communists understood for the first time what it means to be a hated warrior of the previous regime and they saw the destiny that awaits them in it. The fact that the fifty years of communist activity with the goal of condemning everything related to glorious Čiča were insufficient to destroy its last followers has probably given them the hope that, in fifty years, someone will be raising a monument to them too.
>
> (Petrušić, 1992: 12)[7]

The SPO adherents and Serbian diaspora from Canada, the United States and Germany financed the Ravna Gora statue of Mihailović and the Serbian Orthodox Church supported it and its high clergy blessed the statue.[8] They celebrated the statue as the first memorial to Mihailović in Serbia, after the statues the political emigration had erected in Canada, Australia and the United States. In fact, while the statue at Ravna Gora is the most renowned, the first monument was built three months earlier at a private property in Vukošić between Valjevo and Šabac. The younger brother of King Petar II, Tomislav Karađorđević, revealed the statue. Authorities brought the owner of the property and Chetnik sympathiser Đorđe Knežević in for questioning several times. There were also attempts to destroy the monument. According to the article in *Pogledi*, originally published in *Srpske novine* from Chicago, Knežević had to organise a memorial watch. One time when the night watch had relaxed and were drinking alcohol, "the enemies" approached unnoticed and covered the statue with red paint (Topalović, 2016).

In his speech at the 1992 commemoration, Vuk Drašković summarised the interpretation of the Chetnik movement promoted by the SPO and their supporters, particularly after the party changed their position against the wars. For them, the Chetniks became a Yugoslav and democratic movement, gathering all Yugoslav people and, from the St. Sava Congress in Ba in 1944, all relevant political options, naturally excluding the communists. They blamed the Yugoslav communists for starting the civil war, breaking up "the ethnic space of the Serbian people", making alliances with the German, Ustasha and Ballist forces, scheming internationally and representing their anti-Serbdom as antifascism (Drašković, 1992: 22).

For the SPO, the Partisans were the cancer cells that managed to overpower the healthy ones, while Mihailović led "the better part of the nation and state" (Drašković, 1992: 22). Dragoljub Mihailović gathered officers, priests, professors, lawyers, judges, writers, teachers, soldiers who did not violate their oaths and all people who could not be turned away from their nation, morals, tradition and faith by the communist terror and propaganda (Drašković, 1992). Similarly to the DS, Drašković and his adherents saw socialist Yugoslavia as the long, dark night of Serbian history, "a long night without a single star above Serbia" (Drašković, 1992: 22). The last candle extinguished on the night of 17 July 1946, the night of Mihailović's execution.

Not everyone shared the SPO's narrative of the Chetniks as west-oriented Yugoslav democrats. First, as the discourses of national reconciliation and unity became hegemonic in Serbian society in the early 1990s, promoted by the regime, the SUBNOR condemned the Ravna Gora memorial and its supporters as "the traitors and destroyers of the unity of Serbian people in the most difficult moments of the recent history, with unimaginable consequences for Serbia and its citizens" (Stojić, 1992: 8). The strong condemnation of the SPO and their commemorative activities as traitorous arose from the party's criticism about the wars.

The rest of the pro-Chetnik political actors, such as the SRS, were also critical towards the SPO-sponsored commemorations and the memorial as appropriation of the Chetniks and Ravna Gora for purposes of popularity. This division within the memory community characterised by anti-communism and celebration of the Chetniks surfaced in the early 1990s, with the position regarding the wars of the Yugoslav succession becoming a dividing line that also informed the interpretation of the Chetniks. As the SPO embraced the anti-war position, they insisted on the democratic, Yugoslav and West-oriented nature of the Chetnik movement, paradoxically meaning that the Chetniks themselves would have stood against the wars within Yugoslavia. A parallel Chetnik memoryscape existed too, with the SRS as the most notable representatives. These actors were rather fond of Stevan Moljević's plan of homogeneous Serbia, advocating the unification of Serbs in an ethnically homogeneous territory within Yugoslavia. Promoting a similar narrative like the SPO did in the beginning, the SRS, different Chetnik groups and paramilitaries observed the Chetniks and their ideology through the context of the dissolution of Yugoslavia and the wars, seeing it as justifying the wars to bring all Serbs together within the borders of one ethnically homogeneous state.

The division among the Chetnik devotees was visible in the commemorative space. While the SPO-orchestrated gathering at Ravna Gora was becoming increasingly popular, an alternative one began already in the early 1990s, held nearby in the village of Ba where the Chetniks had organised the 1944 St. Sava Congress.[9] In 1992, when thousands of people attended the revealing of the

62 The Milošević era

statue of Mihailović, around 200 people attended the other commemoration, including members of Ravna Gora Movement, not yet registered at that time. In 1995, the Ravna Gora Movement organised the commemoration and had opposition politicians Nikola Milošević, Vojislav Koštunica and Milan St. Protić as speakers. In his speech, Koštunica drew attention to differences between the Chetniks and Partisans, depicting the former as liberating and the latter as enslaving (Janković, 1995: 29). The SPO accused this commemoration of being led by former communists, alluding to Nikola Milošević and Velimir Ilić in 1998,[10] while the "party cell" the SPO referred to accused them of being communists in return (*Srpska reč*, 1998).

As the Chetnik movement was gaining massive popularity, the SPO became one of the main carriers of the Chetnik memory, with close connections to the political emigration. In Ravna Gora, what started as an attempt to put up a memorial plaque in 1990, with a monument of Dragoljub Mihailović added in 1992, grew into an entire memorial complex dedicated to the Chetniks by 2000. With the foundations laid in 1995, the St. George Church was finished in 1998 as a memorial church dedicated to all Serbs who had died in the civil war between 1941 and 1945, usually referred to as the church of reconciliation. The same year, the SPO asphalted the roads leading to the site and started building a large memorial home and completed it in 2000. The parallel commemorations at Ravna Gora continued after 2000 and still take place today, resembling the twofold image of the Chetniks and the divisions sparked by the 1990s wars between the choreographers of Chetnik memory.

In October 2000, three and a half years after the night when the alpinists took down the five-point star from the Assembly's dome, Slobodan Milošević was ousted and his opposition came to power, now in the form of the DOS coalition with Vojislav Koštunica as the president of rump Yugoslavia.

Notes

1. Milošević was president of Serbia from 1990 to 1997 and of Yugoslavia from 1997 to 2000.
2. For instance, in his speech to the Russian delegation in 1995, Milošević called "the Croatian aggression on western Slavonija" an example of contemporary forces openly fighting for the revival of fascism (*Večernje novosti*, 1995: 3).
3. Named House of Flowers (*Kuća cveća*), Tito's mausoleum is a part of the complex of the Museum of Yugoslavia. It was built in 1975 as a winter garden for the nearby residence of the Yugoslav president. Jovanka Broz was buried next to Tito in 2013.
4. The only delegation that laid flowers on Tito's grave in 1992 was the League of Communists – Movement for Yugoslavia (*Savez komunista – Pokret za Jugoslaviju*), the opposition party affiliated with the army that opposed the wars and criticised nationalism. In 1994, the fraction of the party supporting Milošević prevailed and they merged into the Yugoslav United Left (*Jugoslovenska levica, JUL*), led by Mirjana Marković, Milošević's wife and a League member.
5. Vojislav Šešelj established this party before joining the Serbian Renewal Movement with Vuk Drašković. Only a few months later, Šešelj and Drašković split and Šešelj moved on to form the more radical Serbian Chetnik Movement in June 1990, that could not be registered as a political party because of its name. Finally, Šešelj established

the Serbian Radical Party in 1991 by merging Šešelj's adherents with the National Radical Party.
6 Miloslav Samardžić was close to the SPO in 1990 and worked as the editor of their paper, *Srpska reč*. At the time of this manuscript completion in 2019, *Pogledi* still exists online and as a publishing house, while Samardžić, as a member of the far-right *Dveri*, started filming a documentary series about the Second World War and was given media space occasionally where he represented himself as a historian.
7 Dragoljub Mihailović is widely known and usually referred to as Draža, Čiča or Čiča Draža. Čiča means uncle in Serbian.
8 In addition to Vuk Drašković, Artemije Radosavljević, the bishop of Raška and Prizren and Irinej Kovačević, the metropolitan of the diocese of New Gračanica in the United States spoke at the monument revealing ceremony and performed a memorial service.
9 The St. Sava Congress, officially the Congress of the Representatives of Ravna Gora Movement, Political Parties, National Organisations and Institutions and Public Workers (*Kongres predstavnika ravnogorskog pokreta, političkih stranaka, nacionalnih organizacija i ustanova i javnih radnika*), was held between 25 and 28 January 1944 in Ba. As the end of the war and as the loss of all Allied support approached, the Chetniks changed their orientation. The congress can be considered as a turning point when the Chetniks replaced Moljević's homogeneous Greater Serbia with the idea of a democratic and federal Yugoslavia. This is what some of the contemporary Chetnik followers are building upon, such as the SPO, while others still cherish the Moljević's ideas.
10 Velimir Ilić was a member of the SPO until 1997, when he established his own party, *Nova Srbija*.

Bibliography

Bergam, O. (1990) "Mučenik, stradalnik i pobedonosac". *Pogledi*, 26 October.
Bieber, F. (2002) "Nationalist Mobilization and Stories of Serb Suffering: The Kosovo myth from 600th anniversary to the present", *Rethinking History* 6(1): 95–110. DOI: org/10.1080/136425200110112137.
Blic (2017) "Na današnji dan pre 20 godina je skinuta zvezda petokraka, a Beograd je dobio prvog demokratski izabranog gradonačelnika". Available at: www.blic.rs/vesti/beograd/na-danasnji-dan-pre-20-godina-je-skinuta-zvezda-petokraka-a-beograd-je-dobio-prvog/9kdgze4 (accessed 9 August 2018).
Borba (1998a) "Bez prošlosti nema ni budućnosti". 11 May.
Borba (1998b) "Čestitka generala Momčila Perišića". 9 May. Belgrade.
Borba (1998c) "Dan kada je pao fašizam". 9 May.
Čavoški, K. (1990) "Sloboda za sve ljude". *Demokratija*, 9 March. 1st edn. Belgrade.
Ćirić, D. (1990) "Svetski sabor SPO u Beogradu: Nacionalno pomirenje". *Pogledi*, 9 November.
Ćirović, S. (1992) "Srbija na Ravnoj Gori". *Pogledi*, 29 May.
Demokratska stranka (1990) "Pismo o namerama". *Pogledi*, 15 February.
Drašković, V. (1990) "Predlog programa stranke 'Srpska narodna obnova'". *Pogledi*, 20 January.
Drašković, V. (1992) "Kako od poraza do pobede", *Srpska reč*, 25 May.
Drašković, V. (2016) Interview by author.
Đuretić, V. (1990) "Antisrpska 'upotreba' saveznika i revolucije". *NIN*, 6 May.
Janković, J. (1995) "Ponovo na Ravnoj gori". *Pogledi*, 29 May.
Malešević, S. (2002) *Ideology, Legitimacy and the New State: Yugoslavia, Serbia and Croatia*. London: Routledge.

Mandić, V. (1990) "Izlet u mračnu prošlost". *Politika*, 15 May. Belgrade.
Mikuš, M. (2018) *Frontiers of Civil Society: Government and Hegemony in Serbia*. New York: Berghahn Books.
Milošević, N. (1990) "Kritika apsolutne vlasti". *Demokratija*, 9 March. 1st edn. Belgrade.
Ostojić, M. (2014) *Between Justice and Stability: The Politics of War Crimes Prosecutions in Post-Miloševic Serbia*. Surrey: Ashgate.
Otadžbinska centralna uprava srpskog slobodarskog pokreta (1990) "Saopštenje za javnost". *Pogledi*, 4 January.
Pavasović Trošt, T. (2018) "Ruptures and continuities in nationhood narratives: reconstructing the nation through history textbooks in Serbia and Croatia", *Nations and Nationalism* 24(3): 716–740. DOI: org/10.1111/nana.12433.
Pekić, B. (1990) "Moralni aspekti demokratije". *Demokratija*, 9 March. 1st edn. Belgrade.
Petrušić, S. (1992) "Ravna Gora pobediti mora!" *Srpska reč*, 25 May.
Politika (1992) "Titov grob bez počasne straže". 5 May. Belgrade.
Radetić, M. (1992) "Pre osvećenja – rafali". *Večernje novosti*, 14 May. Belgrade.
Radosavljević, S. (1990) "Komunistički zločini: Ja sam zakopavao!" *Pogledi*, 9 November.
Samardžić, M. (1990) "Zašto Srbija treba da bude kraljevina?" *Pogledi*, 15 May.
Srpska reč (1998) "Partijska ćelija u selu Ba". 14 May.
Stefanović, N. (1992) "Drugo buđenje naroda: Vratio se čiča". *Duga*, 23 May.
Stojanović, R. (1990) "O građanskom pomirenju". *Demokratija*, 9 March. 1st edn. Belgrade.
Stojić, J. (1992) "Spomenik generalu Draži – nova deoba srpskog naroda". *Politika*, 13 May. Belgrade.
Subotić, J. (2015) "The Mythologizing of Communist Violence". In Stan, L. and Nedelsky, N. (eds), *Post-Communist Transitional Justice: Lessons from Twenty-Five Years of Experience*. Cambridge: Cambridge University Press, pp. 188–210.
Topalović, D. (2016) "Pokraj druma Šabac – Valjevo: Prvi spomenik Draži u Srbiji". In *Pogledi*. Available at: www.pogledi.rs/pokraj-druma-sabac-valjevo-prvi-spomenik-drazi-u-srbiji/ (accessed 27 September 2018).
Uvalić, M. (2010) *Serbia's Transition: Towards a Better Future*. Basingstoke: Palgrave Macmillan.
Večernje novosti (1995) "Najsvetlije stranice istorije čovečanstva". 9 May. Belgrade.
Vujačić, V. (2004) "Reexamining the 'Serbian Exceptionalism' Thesis", *Berkeley Program in Soviet and Post-Soviet Studies Working Paper Series*: 1–43.
Woodward, S. L. (1995) *Socialist Unemployment: The Political Economy of Yugoslavia, 1945–1990*. Princeton: Princeton University Press.

5 Memory politics in post-Milošević Serbia

Flags of the western Allies, Yugoslavia, the Soviet Union and Serbia, decorated Sava Centar in Belgrade, the venue of the stately academy on the 70th anniversary of the end of the Second World War in 2015. Calling 9 May "the day of strong and victorious Serbia", Prime Minister Aleksandar Vučić (2014–2017, the president of Serbia 2017–) underlined that the Serbian people were the backbone of the war struggle and Serbia sacrificed itself for the grand idea of antifascism, the foundation of Europe (Beljan, 2015; *Večernje novosti*, 2015). Partisan veteran Zdenko Duplančić emphasised that there was only one antifascist movement in Yugoslavia, the one led by Josip Broz Tito and the Communist Party of Yugoslavia. Not directly mentioning the Chetniks, Vučić's interpretation was simply: "Serbs are antifascists", who fought and would always fight fascism (Beljan, 2015; *Večernje novosti*, 2015).

Just five days after the festive academy that celebrated Serbian antifascism and alliance with Russia, the High Court in Belgrade rehabilitated Chetnik leader Dragoljub Mihailović, with the help of Oliver Antić, lawyer and advisor to President Tomislav Nikolić. This was a festive day for many, too. As Antić argued, the day of the positive court decision was a big day for law and justice and a great day not only for the Serbian nation, but also for the honourable Croats, Muslims and Slovenes who fought under Mihailović's command. He said that there should be celebrations in Croatia too (Tanjug, 2015).

While the president and prime minister of Serbia at the time refrained from commenting on the rehabilitation case and never openly celebrated the Chetniks, they never distanced themselves from the Chetniks or their open devotion to and glorification of the Chetnik ideology up until just a few years earlier. When the Serbian Progressive Party (*Srpska napredna stranka*, SNS), a faction of the Serbian Radical Party (SRS), came to power in 2012, they entirely replaced the previously dominant anti-communist paradigm with nationalisation of the Second World War, although these tendencies had surfaced before too. The national lens makes both the Partisans and the Chetniks suitable historical references, reducing the Partisans and their victory to a Serbian achievement, depoliticising, decontextualising and de-Yugoslavising them.

In the period before 2012, the highest state officials similarly abstained from openly glorifying the Chetniks, even though they represented the anti-communist political consensus and played role in the rapid narrative transformation of the immediate post-Milošević period. The exception to this is the Serbian Renewal Movement of Vuk Drašković (SPO), the party whose identity is intertwined with the Chetnik movement. At the same time, the recast images of the Chetniks and the Partisans and the othering and demonisation of socialist Yugoslavia underpinned the overarching mnemonic hegemony in Serbian society.

Three and a half years after the night when Zoran Đinđić took down the five-point star from the local government building, Milošević was overthrown and his opposition came to power in the form of a heterogeneous coalition with Vojislav Koštunica as the president of Serbia. The peaceful transfer of power involved a provisional government and many compromises with the previous regime. Even though the coalition fell apart as soon as 2003, its different members, most notably the Democratic Party (DS), remained prominent in Serbian politics for over a decade.

Although the rule of Milošević was over, the SPS did not disappear from the political sphere in 2000 but was continuously present in the parliament. When *For a European Serbia* coalition, comprised mostly of the 1990s opposition, won the majority in 2008, they formed the government together with the SPS and their electoral allies.[1] Moreover, the DS and SPS presidents signed the Declaration of Political Reconciliation,[2] completing the process of full rehabilitation of the SPS as a legitimate political actor.[3] In the post-Milošević era, the SPS has consistently stood against anti-communist memory politics, presenting themselves as on a mission to preserve the antifascist legacy of Serbia.

The war and its aftermath in the hegemonic narratives

The overthrow of Slobodan Milošević constitutes an immense turning point regarding the memory of the Second World War and socialist Yugoslavia in Serbia, characterised by the transfer of the already existing narratives from the opposition to the sphere of official memory politics and their transformation into the mnemonic hegemony. The positive recasting of the anti-communist forces of the Second World War and the negative reinterpretation of Yugoslavia did not suddenly explode into the public arena, but had existed in the Yugoslav period and the 1990s and found their way into the public sphere. In 2000, these narratives rapidly moved from the micro to macro level of the memory work hierarchy (Conway, 2010).

The heterogeneous coalition that came to power could only agree upon two things: the removal of Milošević and anti-communism, going hand in hand with aiming to make a clear cut with the communist past. This communist past included both socialist Yugoslavia and Milošević's rump Yugoslavia. The political elites narrated and judged Milošević only as a communist, not criticising the nationalist aspects of his ideology and rule (Stojanović, 2011b: 232). In this way, the political opposition that came to power could frame themselves as liberators of Serbia from communism. The interpretation of Milošević as a communist would also "demonstrate that the new authorities were the agents of authentic national values" (Stojanović, 2011b: 232). Since the nationalist discourses persisted after 2000, and numerous other continuities with the 1990s, anti-communism provided an ideal tool for the new government to acquire an identity that would separate them from their predecessors (Stojanović, 2011b: 232). Equally important is that these political parties represented the political opposition of the 1990s, bringing their anti-communism and memory work with them.

Beginning immediately in 2000, the state actors introduced or supported numerous mechanisms to revaluate the communist past. Two parallel processes took place: the removal of Yugoslavia from commemorative practices, official holidays and street names, and the acknowledgement, rehabilitation and compensation for those considered victims of communism. The repression under communism became a very prominent topic, but, as Subotić notes, "almost entire discussion on communist violence in Serbia became a discussion of communist violence against the Chetniks" (2015: 201). The Chetniks became the central theme of the revaluation of the communist past, as both the national anti-fascist movement, a non-communist one, and victims of communist violence.

At the same time, the state actors did not deal with the 1990 wars and their victims, especially the crimes perpetrated by Serbian forces, on such a level. This issue pushed Serbian liberal civil society and "Other Serbia" to engage with the 1990s crimes and their perpetrators, leaving the entire theme of crimes of communism to the anti-communist right (Subotić, 2015: 192, 204).[4] Therefore, no serious discussion or critical assessment of the postwar retribution and other examples of repression and persecution has taken place within the liberal and leftist circles. There are also no campaigns to deal with these issues like there are initiatives to deal with the 1990s.

Postsocialist political elites and the political right hijacked the critical appraisal of Yugoslav state socialism and human rights discourses and transitional justice demands in relation to it. They approached the search for truth and justice for communist crimes with renewed zeal and ambition (Subotić, 2015: 204). The situation in Serbian historiography is remarkably similar. Revisionist historians, small in number but close to the parties that came to power in 2000, made the Second World War and the postwar period their own, securing positions in commissions and other bodies and gaining access to the broader public through media outlets, although the majority of professional historians in Serbia are probably critical of this revision of history.

There are several interconnected and overlapping hegemonic narratives regarding the Second World War, the postwar period and state socialism. One of the initial aims was purging Yugoslavia from the public sphere, including the removal of references to the People's Liberation War. The most visible examples include changing street names and official holidays. The parallel paradigm involves the interpretation of the Second World War as the civil war within the Serbian nation, with the efforts of the postsocialist political elites represented as promoting national reconciliation after the decades of deep division.

The focus of dominant memory politics is on the armed conflict between the Partisans and the Chetniks and their equalisation as a reconciliation of the divided nation, turning the Chetniks into another antifascist movement of Serbia. The apparent equalisation of the opposed sides leads to delegitimisation of the Partisans through the emphasis on postwar retribution and the construction of the narrative of Serbian victimhood under communism. The central theme is the Chetniks and their leader Dragoljub Mihailović as symbols of suffering under communism and its crimes, making the Partisans perpetrators.

The significance of the Second World War for Yugoslav state socialism makes its revision a crucial step in the process of separation from Yugoslavia. The central narrative the political elites promoted after the fall of Milošević was observation of the entire period of the Second World War and decades of state socialism against the backdrop of the postwar trials and executions. The interpretation of a long and complex historical period through a very narrow lens has two goals and outcomes. First, the focus on the revolutionary violence criminalises the People's Liberation Movement and socialist Yugoslavia, either neglecting positive dimensions of this historical experience or representing them as short episodes in the overall negative experience of Yugoslavia for the Serbian nation. Second, the negative image of the Partisans as perpetrators implies a positive reinterpretation of their wartime enemies as both the forces struggling for the protection of Serbian people, unlike the Partisans, and victims of communist terror. Neglecting the wartime activities of these forces – collaboration and mass atrocities – transforms them into innocent victims of communism.

Rehabilitation of anti-communist forces of the Second World War represents the primary objective of Serbia's memory politics. Their revival aims to overturn the value positions: rejecting antifascism and reducing it to communist crimes while socially accepting collaborators as defenders of the nation (Milosavljević, 2013: 227). The concern for victims of communism is symbolic and a façade. In Subotić's words,

> almost the entire effort to uncover communist crimes in post-Milošević Serbia was reduced to the rehabilitation of Chetniks and fascist collaborators as full-fledged patriots who fought the good fight against the onslaught of communist totalitarianism and for the preservation of the Serbian 'golden era', the Kingdom of Yugoslavia.
>
> (2015: 202)

Purging Yugoslavia from the public

The first steps of the institutional level of memory work after the fall of Milošević did not aim at the outright rehabilitation of Second World War defeated forces as victims of communism. The equally important task was eradicating socialist Yugoslavia from the public sphere, particularly the People's Liberation War as the site of "the mythical birth of the communist regime" (Stojanović, 2010: 17). The beginning and the end of the Second World War, earlier celebrated as the beginning of the uprising and the liberation of the country, had to undergo revision. Political actors framed this as separation from the ideologically charged symbols, holidays and names, finally possible after the ultimate defeat of communism symbolised by the overthrow of Milošević.

Changing the street names that used to commemorate the Partisans (and the Red Army) had already taken place in the 1990s, but they accelerated after 2000. Already in 1997, the opposition came to power in Belgrade and replaced street names related to the Partisans and the workers' movement with references to the Kingdom of Yugoslavia (Radović, 2013: 213–216). After 2000, the newly formed Committee for Monuments and Names of Squares and Streets introduced another wave of renaming streets, restoring their pre-communist names. Their goal was cleansing central Belgrade from the People's Liberation War and its postwar Yugoslav past (Radović, 2013: 219–225).

Official holidays could not stay either. In addition to the changes in street names in 1997, 20 October lost its status as the Day of the City of Belgrade, remaining only the Day of the Liberation. In 2002, it was entirely abolished, suiting the view that 20 October did not symbolise the day of liberation but the beginning of the communist occupation. The first mayor of Belgrade in post-Milošević Serbia Milan St. Protić, in office 2000–2001, endorsed this view. As he stated in 2008, "the entrance of the Red Army and Tito's Partisans in Belgrade was the new occupation of Belgrade" (*Blic*, 2008). The October Award, the most prestigious city award, became the Award of the City of Belgrade. It could no longer refer to the month of October as it was moved, together with the official day of the city, to April. There was no official commemoration of the 60th anniversary of the liberation in 2004.

The 2001 Law on State and Other Holidays passed abolished the Day of the Uprising and the Fighter's Day. Abolishing the two holidays symbolising the beginning of the uprising in Serbia and Yugoslavia, the law introduced new public holidays such as Saint Sava's Day, Saint Vitus Day and 15 February as the Serbian Statehood Day (David, 2014). Introducing the law proposal, Minister of Justice Vladan Batić argued that "there were more significant dates in the history of Serbia and those much less ideologised" (*Stenografske beleške sa sedmog vanrednog zasedanja 09.07.2001*, 2001). The change of the value system replaced these "much less ideologised dates" with those that were religious and celebrated Serbia's pre-communist history.

The members of parliament passed the law with 123 out of 183 voting in favour and 39 against. Even the MPs who criticised the law eventually voted for it. This has been the case with all laws regulating the interpretation of the Second World War and Yugoslavia, showing that there are differences in opinion outpowered by a strong consensus in condemnation of state socialism. Only the members of the SPS and other nominally socialist parties opposed the law on holidays and emphasised the positive values 7 July stood for. In the post-Milošević phase, the SPS framed itself as the only party in Serbia preserving the values of the antifascist struggle.

At the international level, the political elites often ignored commemorations of the Second World War and did not send high-ranking representatives during the first post-Milošević decade. In 2005, only the Serbian ambassador to Poland attended the 60th anniversary of the liberation of Auschwitz, without a state delegation from Serbia. The same year, only low-ranking officials attended the Victory Day parade in Moscow. Instead, Prime Minister Vojislav Koštunica (*Demokratska stranka Srbije*, DSS) visited the memorial to the airmen who had defended Belgrade during the 1941 bombing, avoiding any acknowledgement of those who had won the war in Serbia and Yugoslavia (Stojanović, 2011b: 233). President Boris Tadić (DS) chose to lay flowers at a First World War memorial for this occasion. Vuk Drašković, the minister of foreign affairs at the time, went to Ravna Gora, the first and only commemoration sponsored by the Government and the Ministry of Culture in 2005.

National reconciliation: ending the civil war within the Serbian nation

The abolishment of the Day of the Uprising reflects the wider tendency to reduce the complex historical experience of the Second World War on its civil war dimension. The emphasis moved from the beginning of the uprising to the civil war among Serbs, ascribing the guilt for its beginning to the Communist Party of Yugoslavia. The revolutionary dimension of the Partisans' struggle becomes the illegitimate strive of communists to take over the power. In this sense, if socialist Yugoslavia was the "long dark night" of national history that emerged out of the bloody civil war, the return to the pre-communist and pre-Second World War time, the Kingdom of Yugoslavia, is returning to the time when the Serbian nation was not divided. The Yugoslav Army in the Homeland as the army representing the Kingdom of Yugoslavia is a logical reference point for the dominant postsocialist interpretation of Serbian history. According to this view, it was the appearance of the communist-led Partisans that caused the civil war with the Chetniks.

National reconciliation discourse constitutes a significant aspect of the institutional memory work in post-Milošević Serbia and the parliamentary debate on the abolishment of 7 July reflects it. The parliament members spoke of 7 July as the beginning of the tragedy that had lasted for 60 years, representing the shooting of two gendarmes as a Serb killing another Serb for ideological reasons and the day of national shame and disgrace. The shots Španac fired signalled the beginning of "the fratricidal tragedy that we had been suffering from for half of the century and we would suffer from forever" (*Stenografske beleške sa sedmog vanrednog zasedanja 09.07.2001*, 2001). Perceiving themselves as the new beginning, the first parliament members after the overthrow of Milošević saw 5 October 2000 as the end of the civil war in Serbia.

Another legislative step towards national reconciliation was in 2004 when the National Assembly passed the changes to the Law on the Rights of Veterans, Military Invalids and Members of their Families (from now on: The Veteran Law), the act regulating veteran rights and benefits. The law granted the Chetniks, all members of the movement between 17 April 1941 and 15 May 1945, the status of veterans of the People's Liberation War and formally equated them with the Partisans (*Službeni glasnik Republike Srbije*, 2004: 2). Even though the parliament members criticised the law draft heavily, they passed with 113 voting in favour and 24 against, out of 152 present members (Narodna skupština Republike Srbije, 2004b).

The parliamentary proceeding, chaired by Dragoljub Mihailović's grandson Vojislav, extended into a five-day discussion on the Second World War, its aftermath and the nature of the Chetniks. The events of the war were told and retold from different perspectives. The question of whether the Chetniks had fought against the occupation or not became a heated quarrel. To prove their point about the Chetniks' either antifascist or collaborationist nature, the parliament members brought books and documents they quoted, read aloud and showed to others. The parties whose political identity is closely linked to the Chetnik traditions, the SRS, the SPO and their allies, argued about the nature of the Chetnik movement in relation to Serbian nationalism and the wars of the 1990s, competing over the right to be called a Chetnik.

The SPO proposed the law, imagining it as reconciling the Partisans and the Chetniks by not depriving the Partisans of any rights, while giving the Chetniks the same rights that the others had enjoyed for decades. For them, the foundations of these equal rights arose from the armed struggle of both movements against the occupation. The law would show that both movements were antifascist, would prove that the Serbian nation had not been quisling or fascist and would reconcile divided brothers and families. As one of the SPO members explained, if German occupiers had not differentiated between the Partisans and Chetniks when shooting them, there was no reason to differentiate 60 years later. Parties like *Nova Srbija* and Democratic Party of Serbia supported the law. The DS members did not take part in the debate but supported the law proposal. As Milan Marković of the DS explained, both movements had been honourable and had fought against the occupation, but the civil war had also taken place between them. The DS as a party voted for the law because there were "numerous divisions destroying the ethnic tissue of the Serbian nation, so it was necessary to close at least one of them" (Narodna skupština Republike Srbije, 2004b).

Understanding it as a drastic revision of history, the SPS and Social Democratic Party of Serbia (SDP) opposed the law. The SPS emphasised that they were not against pensions for the Chetniks who had fought against the occupation, but they would have to prove it. This SPS amendment made fulfilling pension rights impossible for Chetnik veterans, as the legislation chapter will discuss in detail. Calling the coalition that had come to power in 2000 a Chetnik coalition, Ivica Dačić underlined that the SPS would cherish the memory and tradition of the People's Liberation Movement as long as the party existed (Narodna skupština Republike Srbije, 2004b). The SDP representatives criticised the law as an attempt to give antifascist legitimacy to the Chetnik movement (Narodna skupština Republike Srbije, 2004a). The critiques also addressed the narrative of reconciliation, the SPS claiming that they had reached historical reconciliation by sharing power with their opponents and raised concerns about the deepening of divisions through such laws.

Some parliament members made clear references to the 1990 wars. The SRS, represented by Tomislav Nikolić and Aleksandar Vučić, condemned the law proposal for not truly offering reconciliation, because of a negative portrayal of the Partisans as criminals by law proponents. They connected the idea of "Chetnikhood" (*četništvo*) to the 1990 wars, as depicting a struggle for defending Serbs outside of Serbia and for their unification with their motherland (Narodna skupština Republike Srbije, 2004c). Because of their criticism of the Serbian side in the 1990 wars, Nikolić referred to the SPO party members as "fake Chetniks", explaining:

> I can say with pride: all those who have gone to defend the Serbian lands and people since 1991 can freely call themselves both Chetniks and Partisans. At the frontline, I was called a Chetnik while I was wearing a helmet with a red star. There were no other helmets. If there had been others, if I had been able to choose, I would have chosen, but it was important to save your head back then and the head is not saved by ideology but by equipment. That was the equipment we had, we used it and we did not mind it and no one in Krajina laughed at us for wearing red star helmets, we were laughed at around Belgrade. Those who hid under their wives' skirts and turned off lights when the mobilisation call came, they used to say: "What kind of Chetniks are you, wearing a red star?" Those were the best Chetniks.
> (Narodna skupština Republike Srbije, 2004c)

Imaginations of the Chetniks

The Veteran Law changes represent the official confirmation of the long process of positive revaluation of the Chetniks, including their reinvention as antifascists and celebration of their anti-communism. While it is legitimate to interpret the Chetniks of Dragoljub Mihailović as a resistance movement during the first months of the Second World War in Yugoslavia, the institutional level of memory work and a part of Serbian historiography represent them as unconditionally antifascist throughout the war. Even though they never referred to themselves as antifascists, the Chetniks acquired this epithet, as declared veterans of the People's Liberation War, in which they had only taken place as the Partisans' adversaries. Ignoring their collaboration and crimes and adding antifascism makes the Chetniks "the ideal ancestors" who accidentally found themselves on the defeated side (Stojanović, 2011a: 248). In the post-2000 period, different media with state support, including textbooks, television series and museums, promoted this image.

Chetnik anti-communism constitutes another vital aspect of their status as a positive historical reference. From autumn 1941, anti-communism motivated the Chetniks to dedicate their greatest efforts to fighting the Partisans, while their position against the occupation remained mostly nominal. It was anti-communism that drove the Chetniks into collaboration with the occupation and quisling forces. Because of it, some Chetnik units also committed numerous atrocities against alleged communists and "Partisan villages" in Serbia, while the atrocities against the non-Serb civilian population were motivated by the ideas of an ethnically homogeneous Serbian territory or represented retaliatory measures for crimes against the Serbian population. Numerous atrocities outside of Serbia contribute to the negative image of the Chetniks in Bosnia and Herzegovina today, equally relevant as the Chetnik revival of the 1990s. At the same time, they officially represented the Yugoslav royal government and had Allied support until 1943. If the Chetnik collaboration and crimes are put aside, the Yugoslav Army in the Homeland becomes a legitimate king's army and a national movement fighting against foreign occupation and for the protection of Serbian people as part of the international anti-Hitler coalition.

The hegemonic discourses, a large segment of which stems from the 1980s and 1990s, underline and overstate the popular support for the Chetniks, especially by the ordinary people in Serbia proper. In addition to the Chetnik memoirs that celebrate anecdotes and images of Draža among the people, as Dragoljub Mihailović is usually referred to, they became a commonplace in the public representations of the Chetniks. As opposed to that, the victory of the Partisans becomes linked to the decisive support of the Red Army in 1944. Blinded by the ideology foreign to the Serbian nation, the Partisans couldn't, thus, achieve wide popular support and they did not care about the people either when their attacks on German occupation forces caused massive reprisals. According to this view, the end of the Second World War represents defeat for the divided Serbian nation. As historian Čedomir Antić summarises in the preface to Zvonimir Vučković's war memoirs:

> The last war for the defence of Serbdom, the Second World War, ended with the defeat, incorporating Serbia into a federation where the national and state interests of her people were systematically neglected and in a political order that had been advocated by a societal, social, and moral minority before 1941.
>
> (Vučković, 2015: vii)

While the above outlined discourses constitute the hegemonic memory culture, there is more than one vision of the Chetniks, as the 1990s division among their sympathisers shows. The SRS kept their view of the Chetniks as anti-Western and the idea of the Chetnikhood as the struggle for the unification of all Serbs within one state border. For the SPO, the Chetniks were still pro-Western, European, Yugoslav and democratic, and if 27 March 1941 had not happened and Yugoslavia had stayed neutral in the Second World War, "we would still live in Yugoslavia today, the 1990s wars would have not happened, and this would have been a decent European constitutional parliamentary monarchy" (Drašković, 2016). While Šešelj and his followers criticised Drašković for being a bogus Chetnik, the SPO argues that nationalist movements hid behind Mihailović and his pro-Western guerrillas to oppose the strategy of Serbia's accession to the European Union and NATO.

New pro-Chetnik mnemonic agents emerged or grew larger in post-Milošević Serbia. A fraction expelled from the SPO established a political party in 2017, entitled the Movement for the Renewal of the Kingdom of Serbia (*Pokret obnove Kraljevine Srbije, POKS*). The POKS represents a more radical wing of the SPO that promotes going back to the values and goals of the party of the early 1990s, namely the reestablishment of the monarchy in Serbia. The movement formed on the anniversary of the death of Dragoljub Mihailović and commemorates the Chetniks as the king's army, while promoting anti-communism. Vojislav Mihailović is affiliated with them.

Another group, closer to the SRS in the Chetnik image they cherish, is the Ravna Gora Movement. This movement had been organising the Ravna Gora commemorations parallel to the SPO gathering already in the 1990s, with the support from prominent politically active intellectuals. In the post-Milošević period, they grew into a large militant movement, involving Serbian veterans of the 1990s wars with offices in both Serbia and Republika Srpska. Their image of the Chetniks intertwines with a pro-Russian position, so they took part in the conflict in western Ukraine wearing Chetnik uniforms, and the annual commemorations of Mihailović's death involve the memorial service for Nikolai II and the Romanov family.

The heterogeneous nature of Chetnik memory does not resonate outside of Serbia. While the memory and memorialisation of the Chetniks represents a multifaceted phenomenon in Serbia, the external critiques, particularly from Bosnia and Croatia, are based on the sense of one single Chetnik rehabilitation narrative. This became particularly evident during the process of rehabilitation of Dragoljub Mihailović that caused outrage among political actors in these countries. I will return to this issue in the chapter dedicated to the court case.

As the example of the Ravna Gora Movement illuminates, the Serbian-Russian relations do not only inform the restoration of antifascist commemorations such as the state-sponsored Victory Day celebrations. Similarly, the mnemonic aspect of Serbia's European Union accession does not only relate to the narrative of antifascism as a European value rooted in the EU foundations, as Serbian political elites and liberal civil society often emphasise – the first to justify the revival of Partisan commemorations and the latter to fight revisionist memory politics that could, in their opinion, harm the EU accession (*Blic*, 2015). Political elites and revisionist historians utilise the Europeanisation discourses to rationalise revisionism in Serbia, idolising the antitotalitarian resolutions and EU memory politics. As Ana Milošević and Heleen Tourquet demonstrate, the SPO has repeatedly proposed a resolution on condemnation of crimes of communism and the establishment of a remembrance day for victims of totalitarian regimes since 2009. Although without success, these attempts show how the EU resolutions "are utilised in attempts to rehabilitate the Chetniks" (Milošević and Touquet, 2018: 12). The SPO parliament members have also lobbied for the opening of state security files:

> It is a very important test for the whole society, for all ruling structures, and it is one of the mandatory conditions for the path of Serbia towards the EU. I think the law should have been passed a long time ago, as it was done in Germany after the unification when the files of the secret police were

opened. It is obvious that the unreformed security services here are still preventing such a law to be passed.

(A. Čotrić, 2016)

The position towards the EU accession and the Russian Federation is another demarcation line of the contemporary Chetnik adherents, in a similar way to the stance towards the 1990 wars, both informing the interpretation of the Chetniks. Hence, while the Ravna Gora Movement members fought in Ukraine and commemorated Mihailović together with the Romanovs, Aleksandar Čotrić from the SPO explained that "if General Mihailović was alive today, he would without a doubt be for the European Union and strategic partnership with the United States of America. He would also certainly not be in favor of the Eurasian Union" (A. Čotrić, 2016).

While being their loudest and most prominent proponents who seek to represent themselves as political successors of Mihailović's Chetniks, it was not the SPO and SRS alone that have rehabilitated the Chetniks in Serbia. Their positive reinterpretation as central to state-sanctioned memory politics is the outcome of the broad anti-communist consensus of the post-Milošević political elites. This consensus is evident in the parliamentary debates that end with the clear majority voting in favour of legislation that regulates the interpretation of the past. The DS and the parties affiliated with it either initiated or supported all mechanisms of memory politics introduced since 2000, voting in favour of all legislation proposals.

The Chetniks televised

Television has contributed to spreading the positive image of the Chetnik movement. The public broadcaster RTS (*Radio televizija Srbije*) has aired two series in the post-2000 period with a focus on the Chetniks: *Ravnogorska čitanka* in 2002 and *Ravna Gora* in 2013. Following the Second World War in Serbia with the focus on the Chetniks as a resistance movement of honest and respectable Serbian men, the documentary series *Ravnogorska čitanka* emerged from the book edited by series author Uglješa Krstić. His motivation was to contribute to the decommunisation process in Serbia, claiming that "Germans had their denazification and we should have the decommunisation" (Grujičić, 2002). The series involved Bojan Dimitrijević as an expert consultant, a professional historian, DS member and the advisor of the minister of defence at the time. Political actors greeted the series. Announcing the draft of legislation on rehabilitation of all victims of persecution and repression, including the Chetniks, Vladan Batić (Minister of Justice 2001–2003) argued that the series could contribute to reconciliation:

Enough is enough with media politisation, distortion of reality, and falsification of history. In this light, I think it is good that such a series is aired, made from a professional and unbiased angle. It is about time to finally put an end to the half century long divisions.

(Grujičić, 2002)

Provoking more reactions, the *Ravna Gora* feature series premiered at prime time in November 2013. The series followed the beginning of the Second World War through the story of a Serbian family with one son in the Partisans and the other in the Chetniks. The idea came from former RTS director Aleksandar Tijanić and the RTS, the public broadcaster funded by the citizens of Serbia and state subsidies, partly financed the series. According to the series author Radoš Bajić, the RTS contributed €90,000 to the budget of every episode that was 120,000 in total (*Večernje novosti*, 2014).

Reproducing the narrative of Yugoslavia as the dark night of national history and promoting his series as showing the light, Bajić claimed that "seventy years after the Second World War, we were still sitting in the dark, with an occasional spark of truth that sporadically shined on us" (Milojković, 2013). The RTS promoted the series as calming the passion and leading to national reconciliation, while Bajić advertised it as a "demystification of ideological fallacies and historical injustice" (Ivanović and Rivljanin, 2013).

As soon as the first trailer appeared, however, so did the critiques that the series served the rehabilitation of the Chetniks. Perceiving the critiques as attacking not only the series, but also the Chetnik movement in general, Vuk Drašković responded: "There is no more Tito, nor Draža, nor Yugoslavia, but the communist condemnation of *Ravna Gora* and its commander still lives in that dead country, and the hunt is led by the still living Partisans and their ideological progeny" (*Telegraf*, 2013). As an expert consultant for *Ravna Gora*, Momčilo Pavlović from the Institute for Contemporary History also defended the series, claiming it represented a balanced view on the war (Đorđević, 2013). More than two million people watched the first episode.

Besides historians and film critics, the series actors continuously offered their interpretations of the Second World War. For instance, Nebojša Glogovac who played the role of Mihailović, provided elaborate explanations about the Chetnik movement and the war. He argued that "Draža only wanted to defend his country, while everything else is propaganda", calling the civil war between the Chetniks and the Partisans a fratricidal war (*Vesti online*, 2013).

While *Ravna Gora* does not glorify the Chetniks, it certainly shows them and Dragoljub Mihailović in a very positive light, avoiding controversial issues. However, the series only engaged with the beginning of the war and its framework does not encompass the Chetnik collaboration and atrocities as well as the conflict with the Partisans that emerged later. The series involves many factual inaccuracies and mistakes. Although it does not belong to the documentary genre, unlike *Ravnogorska čitanka*, it allows the author to represent his take on the war, the involvement of historians and institutes giving it more credibility as a historical project.

With the differences in genre, the television series had many common characteristics. They both promoted the idea of national reconciliation through the revision of the Second World War interpretation that had been dominant in socialist Yugoslavia and the turn to the Chetnik tradition. They reflected the post-Yugoslav memory culture in Serbia, but they also represented the tool of the state-sanctioned memory politics, partially financed by the state by being aired on the public service broadcaster, providing them with a large audience.

War musealisation

The revised permanent museum exhibitions in Serbia usually do not resemble the typical postsocialist museums, such as the House of Terror in Budapest, that focus on the victimhood of the nation under communism and are based on antitotalitarian paradigm with the representation of communism as the more negative totalitarian regime. The only example of a museal representation of history that entails thematic and visual similarities with the museums in Budapest, Vilnius or Riga is the travelling and only partially state-funded exhibition *In the Name of the People! Political Repression in Serbia 1944–1953* (*U ime naroda! Politička represija u Srbiji 1944–1953*) that I will discuss later. In some cases, exhibitions emphasise the postwar retribution as a crucial episode in the history of the city.[5]

The publicly funded museums in Serbia usually do not strive for emotional reaction from the visitors and they depict history in a rather static, factual and textbook-style way, often depoliticising historical events, actors and their conflicts, such as the Chetniks and Partisans. For instance, the memorial museum of the former concentration camp in Niš, in its most recent exhibition that was opened in 2013, dedicates space to the suffering of members of the Yugoslav Army in the Homeland, indirectly representing them as an antifascist movement persecuted by the German occupation. Exhibitions usually entail black and white photographs and long texts, often accompanied by objects of the military provenance. In some museums, like in Kraljevo, the objects affiliated with the Chetniks and Partisans stand next to each other without additional information or differentiation between the movements.

The Museum of Užička Republika in Užice and the Military Museum in Belgrade are interesting examples of the transformation of formerly Yugoslav museums. When changing the previous permanent exhibition on the Second World War, neither of the two museums altered the Yugoslav representations of the Partisans. They simply added the Chetniks to that narrative. In the case of Užice, where the new exhibition opened for the 75th anniversary of Užička republika in 2016, the panel that introduces the exhibition represents the Partisans and Chetniks as two resistance movements. Their weapons and other objects stand next to each other. On the other hand, the exhibition does not leave out the Chetnik crimes and collaboration, including their fights against the Partisans, while openly glorifying the Partisans. Because of that, the exhibition is thematically very close to the museal representations of the socialist period.

The exhibition, however, depoliticises the conflict between the two movements and represents it as a fratricidal war within the Serbian nation provoked by the disagreement about mobilisation and money. Curator of the exhibition,

Radivoj Papić, explains that it was anyway irrelevant which movement one joined as they had both been positioned against the occupation:

> They both emphasised patriotism. People applied for patriotic reasons. At the beginning, people used to say that it did not matter whether you are in Chetniks or Partisans but whether you are for fighting the occupiers. They even had two desks in the same place and then people freely expressed who would go to the Partisans and who to the Chetniks, so one brother would opt for the Partisans and the other for the Chetniks, it did not matter, whatever. The split came later so it happened that a brother would shoot his brother, a son his father, and so on, because they found themselves in different positions.
> (Papić, 2016)

The Military Museum in Belgrade is a fascinating example of the postsocialist musealisation of the Second World War in Serbia. Tito opened the museum as it is today, within the walls of the Belgrade fortress, on 20 October 1961. Most of the current permanent exhibition comes from the socialist period and represents the regional history from the early modern times through the lens of class war. An entire floor is dedicated to the Second World War. The yellowed text panels, that used to be white, accompany the diverse photographs, objects and artwork about the People's Liberation War.

The museum revised the segment of the permanent exhibition that deals with the interwar period and the Second World War and opened it on 8 May 2010, for the celebration of Victory Day. The interventions particularly concern the Chetniks, simply added to the exhibition without significant changes to the original 1961 People's Liberation War exhibition. The exhibition authors literally squeezed in the Chetniks wherever they found space, adding the text inscribed on very bright white and thus visually different panels to the old and yellowed ones. In the middle of the exhibition that follows the narrative of the Partisans' struggle, the visitors suddenly find the photographs of Dragoljub Mihailović with British officers and the text panels about the Chetniks saving the lives of American pilots, all printed on bright new paper. Placing the numerous objects related to the Chetniks and Partisans next to each other throughout the whole exhibition, such as the typical wartime garb worn by their members, represents them as two Serbian resistance movements. The exhibition includes personal objects that had belonged to Mihailović and Tito respectively, before 2010 not available to the public.[6]

The addition of the Chetniks to the exhibitions that had the Partisans in focus could be a result of the lack of finances for the thorough revision of permanent exhibitions in Serbian museums that would change the narrative of the war and the postwar time more drastically. The example of Šabac, with significant funding invested in the new permanent exhibition that focuses on the postwar purges and which is anti-communist- and Chetnik-oriented, contributes to this argument. Nevertheless, the museums in Užice and Belgrade also reflect the wider paradigm in the musealisation of the Second World War in Serbia that equalises the Chetniks and Partisans. Such representations do not promote anti-communism outright but are rather based on the nationalisation of the past. According to this view, as Radivoj Papić explained above, it only matters that the Serbs constituted a majority within both movements and that these were both resistance to the occupation. Placing them shoulder to shoulder in a museum is an aspect of their reconciliation.

The Chetniks as victims of communism

Victimhood under communism is equally as relevant as Chetniks' representation as a resistance movement and their official equalisation with the Partisans. Focusing on the postwar executions and repression, the Serbian political elites dedicated enormous efforts to the acknowledgement and compensation for those considered victims of communism. These efforts did not go far beyond a rhetorical level, although mnemonic agents narrate their actions as dedicated to all innocent victims of communism and righting the past wrongs. The Chetniks take the highest position within the hierarchy of victims of communism, Dragoljub Mihailović standing as the central symbol of the movement's victimhood.

The focus on the victimhood of the Chetniks informs the image of the Partisans, as it turns the apparent equalisation into the delegitimisation of the Partisans as perpetrators. The dominant discourses blame the Partisans for starting the civil war on 7 July 1941, when a Partisan killed two Serbian gendarmes, and declare them perpetrators for settling accounts with collaborators and enemies at the end of the war, while their wartime efforts against the occupation are neglected. In other words, the narrow lens of the postwar repression equally affects the Partisans and their enemies. Neglecting the wartime activities of the military and political forces, such as the Chetniks and their collaboration and atrocities, constructs them as positive historical references and innocent victims of communism, just as it constructs the Partisans as exclusively perpetrators. The proponents of the Chetnik rehabilitation, while promoting different interpretations of the Chetniks, all agree on the image of the Partisans as perpetrators who are to blame for the civil war, importing the foreign ideology and occupying the Serbian nation for decades.

The narrative of the forbidden and repressed history underlines public discussion on the postwar retribution or the positive war stories involving the Chetniks, regardless of whether it takes place in the parliament, media or museums, as opening the topic after decades of suppression. As the previous chapters illuminate, the debates on these issues existed in the 1980s, including the Chetniks as a resistance movement, the Partisans' violence and repression in socialist Yugoslavia. In the 1990s, the political magazines such as *Pogledi* or *NIN* published long-running feuilletons dedicated to the postwar executions in Serbia. The same political parties whose members claim it was finally possible to discuss the suppressed history after 2000 had themselves addressed this history publicly in the late 1980s and early 1990s.

Images of the dark past

As formerly vernacular memories of crimes of communism and the defeated forces of the Second World War transformed into hegemonic discourses in 2000, the form of their media representation changed from the textual forms to the more complex and resource-demanding media. While these narratives had predominantly reached the public as newspaper articles, feuilletons and books before, they became a topic of numerous television series shown and co-financed by the public broadcaster after 2000, such as *Ravnogorska čitanka* and *Ravna Gora*. The anti-communist political climate and consensus enabled their production.

The first television project dealing with the postwar period was not dedicated specifically to the Chetniks, but they were its vital part. Historian Dušan T. Bataković conceptualised the series under the title *Red Age: Crimes of Communists in Serbia and Montenegro 1944–1947* (*Crveno doba*). The documentary series was the result of a project led by Bataković and his non-governmental organisation, the Council for Democratic Changes (*Veće za demokratske promene*), a non-state initiative of gathering archival sources and witness statements on the postwar period that involved young historians and journalists. Bataković prepared and ran the project for ten years, before he found sponsors, formed the research team and composed the list of interviewees. It was also him who eventually convinced the RTS to air it. While his associates received some financial compensation, Bataković worked for free (Bataković, 2005).

The project coordinator was historian Slobodan G. Marković, later the president of the State Commission for Secret Graves and one of the most prominent proponents of rehabilitation of Dragoljub Mihailović.[7] The project was about confronting the totalitarian legacies of communism and emphasised the destruction of Serbian elites by the Partisans. For Marković, understanding the causes for the fall of communist Yugoslavia should arise from understanding its formation and "the entire process of the communist tyranny from its institutionalisation in 1944 to its metastasis during the 1990s" (Otašević, 2001).

The project and the series had the goal of reconciling the two antifascist movements (Otašević, 2001). The Chetniks and their defeat featured prominently in the series that arose from the project, involving interviews with Chetnik veterans. The series depicted the Chetniks in a significantly more positive light than the Partisans and their defeat as a tragedy. The 1944 liberation from Serbia was a negative event that brought, as Bataković narrates in the series, "terror, terror and only terror".

The only museum exhibition related to the Second World War with resources and a multimedia approach is the one dedicated to communist terror, while the possibilities of other museums, that are publicly funded, remain limited. In April 2014, the *U ime naroda* exhibition opened in the Historical Museum of Serbia in Belgrade. Initiated by historian Srđan Cvetković as the main author, the exhibition had the support of the Ministry of Culture. It was the first museum representation of the immediate aftermath of the Second World War in Serbia. The Historical

Museum announced it as the first museum exhibition "about the decades long taboo topics: executions and trials against 'people's enemies', Goli Otok and other camps, forced collectivisation, elections, the cult of personality and political culture between 1944 and 1953" (Nikolić, 2014). The concrete five-point star, that had been taken down by alpinists and given to Zoran Đinđić in 1997, awaited the visitors at the entrance.

The media, including the RTS, advertised *U ime naroda* as revealing the hidden dark side of Serbian history. Describing the goals of the exhibition as "making people in the 21st century feel the atmosphere of the repression, implemented first as the imitation of Stalinism and then as totalitarianism in Tito's style", Cvetković wanted to turn off electricity at the square in front of the museum, so that the darkness would illustrate the atmosphere of the postwar terror (Lopušina, 2013). The basic idea of the exhibition project was the representation of the entire period of state socialism as dark through focusing on its most repressive part. The exhibition author intended to show the real nature of Yugoslav socialism:

> What is, in my opinion, deliberately forgotten is that the executions represented only the entrance to a repressive system that would last for decades. It was not something isolated like in France, that it happened and had its beginning and its end, short, a few months, and then some parties and coalitions came so a democratic system and rule of law were established. No. Here, it was the beginning of a lawless state, a state that considered the judiciary as the arm of the party.... It was the introduction to the massive system of repression through arrests and political trials, this time with prison sentences, rarely death sentences, of many young people, many peasants because of the Stalinisation of villages, many party opponents ... different social groups that bothered the ruling oligarchy. The introduction to a society of all-encompassing fear, denunciators, the pioneer improvisation, a society that carried batons for the Day of the Youth. A concept of North Korea. Yes, really! We are looking at North Korea now wondering but that is the image of this country 50 years ago, only 50 years ago. Later, it all became moderate because of foreign pressures, liberalised, but never until the end, never completely. That repression was always hanging upon people, like the Sword of Damocles, that violence, the fear, it crept into people, it was whispered in the 1970s and even in the 1980s.
>
> (Cvetković, 2016)

The exhibition was about "ordinary, anonymous people, peasants, priests, sportsmen, professors" whose destinies were no less relevant than those of the more prominent people (Popadić, 2014). However, following the dominant paradigm of a selective approach to history and exonerating collaborators as innocent victims of communism, the exhibition poster, a photo mosaic with the portraits of victims of revolutionary terror, also showed Milan Aćimović, the Head of the Commissary Government and Minister of Interior in Milan Nedić's government. More importantly, Dragoljub Mihailović had the central spot in the exhibition, with his personal belongings exhibited in a prison cell that was supposed to resemble where he had spent the last days of his life. While the belongings Mihailović had had with him during the capture had already been shown in two museums before, the media reported it in a sensationalist manner, as if they had never been presented to the public.[8]

Having reached a large audience in Serbia, the television series and the exhibition represent only the examples of the emphasis on the Chetniks and Dragoljub Mihailović within the discourse of the Serbian victimhood under communism. This is, however, evident in other spheres too. For instance, the history sections of most bookshops in Serbia are dominated by the biographies of Dragoljub Mihailović and books about his victimisation through the 1946 trial. While this genre emerged in the 1990s, as the Chetnik memoirs were increasingly published in Serbia, there was an explosion of books about the Chetniks and Dragoljub Mihailović in the post-Milošević period, now also published by the state-supported publishers.

The state-sanctioned measures directed at dealing with the communist past, namely the fact-finding commissions and mechanism of judicial rehabilitation, are particularly relevant for making the Chetnik victimisation the dominant interpretation of the movement, in addition to their reframing as an antifascist force. As the following chapters will demonstrate, the centrality of the Chetniks in the narrative of victimhood under communism is evident in the establishment of a fact-finding commission with a task to look for Mihailović's remains. Judicial rehabilitation of Mihailović in 2015 made his victim status official. Moreover, the interest in media actors in these issues, the public attention and both positive and negative reactions further contributed to the omnipresence of the narrative of the Chetnik victimhood in the public sphere in Serbia.

Notes

1 *Za evropsku Srbiju*. The coalition was led by Boris Tadić of the DS and comprised of the DS, SPO, the League of Social Democrats of Vojvodina (*Liga socijaldemokrata Vojvodine*, LSV), G17 and others.
2 Peščanik (n.d.) *Deklaracija o pomirenju DS i SPS*. Available at: https://pescanik.net/deklaracija-o-pomirenju-ds-i-sps/ (accessed 9 August 2018).
3 Kosovo declared independence in February of the same year. Moreover, also in 2008, one of the largest political parties, the Serbian Radical Party (SRS), split when a fraction led by Tomislav Nikolić stepped out to form the SNS. The SNS has dominated Serbian politics since 2012, in the coalition with the SPS, which became the second most influential political party in Serbia.
4 "Other Serbia" *(druga Srbija)* or "Civil Serbia" (*građanska Srbija*) emerged in the 1990s among the cultural, intellectual and political elites and middle classes gathered around anti-war, anti-nationalist, liberal and pro-EU politics (Mikuš, 2018).
5 The permanent exhibition in the National Museum in Šabac ends with an oversized photograph of the railway bridge where the locals had been shot for collaboration after the liberation of the city in 1944, with their names inscribed, while leaving out the wartime circumstances. The exhibition does not mention the local concentration camp and suffering of the local population during the occupation, nor does it refer to the

disappearance of the Jewish population, both local and those that had arrived to Šabac in 1940 as a part of the Kladovo transport.
6 The exhibition ends with the 1990s wars and NATO bombing of Yugoslavia, adding of which is another paradigm of the postsocialist museum transformation across the post-Yugoslav space.
7 Both Marković and Bataković had affiliation with the Karađorđević family, as members of the Crown and Privy Council respectively.
8 The exhibition program also involved weekly lectures, discussions and film screenings that took place in a room referred to as "the Partisan courtroom" and they attracted a large audience.

Bibliography

Bataković, D. T. (2005) "Prisvajanje autorstva". Available at: www.vreme.com/cms/view.php?id=405163 (accessed 22 August 2018).

Beljan, M. (2015) "Vučić: Slavimo Dan pobede jake, pobedničke Srbije". Available at: www.blic.rs/vesti/politika/vucic-slavimo-dan-pobede-jake-pobednicke-srbije/h9r6sxl (accessed 15 December 2018).

Blic (2008) "Na 20. oktobar podsećaju škola i tri ulice". Available at: www.blic.rs/vesti/beograd/na-20-oktobar-podsecaju-skola-i-tri-ulice/ctw823s (accessed 17 March 2018).

Blic (2015) "Helsinški odbor: Srbija odustala od antifašizma". Available at: www.blic.rs/vesti/politika/helsinski-odbor-srbija-odustala-od-antifasizma/mdg96dx (accessed 29 September 2018).

Conway, B. (2010) Commemoration and Bloody Sunday: Pathways of Memory. Palgrave Macmillan Memory Studies. Basingstoke: Palgrave Macmillan UK.

Čotrić, Aleksandar (2016) "Draža bi bio za Evropsku uniju". Available at: www.danas.rs/dijalog/licni-stavovi/draza-bi-bio-za-evropsku-uniju/ (accessed 29 September 2018).

Čotrić, A. (2016) Interview by author.

Cvetković, S. (2016) Interview by author.

David, L. (2014) "Impression Management of a Contested Past: Serbia's Evolving National Calendar", Memory Studies 7(4): 472–483. DOI: org/10.1177/1750698014537670.

Đorđević, K. (2013) "'Ravnu goru' gledalo više od dva miliona gledalaca". Available at: www.politika.rs/scc/clanak/275534/Ravnu-goru-gledalo-vise-od-dva-miliona-gledalaca (accessed 29 September 2018).

Drašković, V. (2016) Interview by author.

Grujičić, N. (2002) "Četnici i početnici". Vreme, 5 February. Belgrade. Available at: www.vreme.com/cms/view.php?id=313416 (accessed 20 August 2018).

Ivanović, D. and Rivljanin, R. (2013) "Radoš Bajić: Izviniće mi se za sve uvrede zbog 'Ravne gore'". Available at: www.blic.rs/vesti/drustvo/rados-bajic-izvinice-mi-se-za-sve-uvrede-zbog-ravne-gore/6fghv97 (accessed 30 September 2018).

Lopušina, M. (2013) "Srđan Cvetković: Osvetljavamo mračne godine Srbije". Available at: www.novosti.rs/vesti/naslovna/drustvo/aktuelno.290.html:458560-Srdjan-Cvetkovic-Osvetljavamo-mracne-godine-Srbije (accessed 2 October 2018).

Mikuš, M. (2018) Frontiers of Civil Society: Government and Hegemony in Serbia. New York: Berghahn Books.

Milojković, S. (2013) "Radoš Bajić: Komunizam se nigde nije primio kao kod nas u Srbiji". Available at: www.blic.rs/kultura/vesti/rados-bajic-komunizam-se-nigde-nije-primio-kao-kod-nas-u-srbiji/xnpew98 (accessed 30 September 2018).

Milosavljević, O. (2013) "Geschichtsrevisionismus und der Zweite Weltkrieg". In Tomić, Đ., Zschächner, R., Puškarević, M. and Schneider, A. (eds), Mythos Partizan. (Dis)

Kontinuitäten der jugoslawischen Linken: Geschichte, Erinnerungen und Perspektiven. Münster: Unrast Verlag, pp. 222–234.

Milošević, A. and Touquet, H. (2018) "Unintended Consequences: The EU Memory Framework and the Politics of Memory in Serbia and Croatia", *Southeast European and Black Sea Studies* 18(3): 381–399. DOI: org/10.1080/14683857.2018.1489614.

Narodna skupština Republike Srbije (2004a) "Privremene stenografske beleške, sedma sednica drugog redovnog zasedanja, 3. dan rada". Available at: www.otvoreniparlament.rs/transkript/6172?page=1#govor-897964 (accessed 29 August 2018).

Narodna skupština Republike Srbije (2004b) "Privremene stenografske beleške, sedma sednica drugog redovnog zasedanja, 5. dan rada". Available at: www.otvoreniparlament.rs/transkript/6174?tagId=60655ž (accessed 9 October 2018).

Narodna skupština Republike Srbije (2004c) "Privremene stenografske beleške, šesta sednica drugog redovnog zasedanja". Available at: www.otvoreniparlament.rs/transkript/6170?tagId=60685 (accessed 29 August 2018).

Nikolić, A. (2014) "Burno na otvaranju izložbe 'U ime naroda': Skojevci, policija, poklici 'bando crvena' ..." Available at: www.blic.rs/kultura/vesti/burno-na-otvaranju-izlozbe-u-ime-naroda-skojevci-policija-poklici-bando-crvena/7rvvpdh (accessed 2 October 2018).

Otašević, A. (2001) "Sudbina pobeđenih". *NIN*, 25 October. Available at: www.nin.co.rs/2001-10/25/20379.html (accessed 22 August 2018).

Papić, R. (2016) Interview by author.

Popadić, A. (2014) "Slike mračne prošlosti". Available at: www.novosti.rs/vesti/kultura.71.html:486961-Slike-mracne-proslosti (accessed 2 October 2018).

Radović, S. (2013) *Grad kao tekst*. Belgrade: Biblioteka XX vek.

Službeni glasnik Republike Srbije (2004) Zakon o pravima boraca, vojnih invalida i njihovih porodica 2004.

Stenografske beleške sa sedmog vanrednog zasedanja 09.07.2001. (2001). Available at: www.otvoreniparlament.rs/2001/07/09/370689 (accessed 7 October 2017).

Stojanović, D. (2010) "U ogledalu drugih". In Dimitrijević, V. (ed.), *Novosti iz prošlosti: Znanje, neznanje, upotreba i zloupotreba istorije*. Belgrade: Belgrade Center for Human Rights, pp. 13–33.

Stojanović, D. (2011a) "Revisions of the Second World War History in Serbia". In Ramet, S. P. and Listhaug, O. (eds), *Serbia and the Serbs in World War Two*. Basingstoke: Palgrave Macmillan UK, pp. 247–364.

Stojanović, D. (2011b) "Value Changes in the Interpretations of History in Serbia". In Ramet, S. P., Listhaug, O. and Dulić, D. (eds), *Civic and Uncivil Values: Serbia in the Post-Milošević Era*. Budapest: Central European University Press, pp. 221–241.

Subotić, J. (2015) "The Mythologizing of Communist Violence". In Stan, L. and Nedelsky, N. (eds), *Post-Communist Transitional Justice: Lessons from Twenty-Five Years of Experience*. Cambridge: Cambridge University Press, pp. 188–210.

Tanjug (2015) "Nikolićev savetnik: Rehabilitaciju Draže treba da slave i u Hrvatskoj". Available at: www.blic.rs/vesti/drustvo/nikolicev-savetnik-rehabilitaciju-draze-treba-da-slave-i-u-hrvatskoj/mgxpzng (accessed 11 July 2019).

Telegraf (2013) "Drašković: Srpski partizani razapinju Dražu i 'Ravnu goru'". Available at: www.telegraf.rs/vesti/880135-draskovic-srpski-partizani-i-dalje-razapinju-drazu-i-ravnu-goru (accessed 30 September 2018).

Večernje novosti (2014) "Ravna gora i dalje čeka istinu". Available at: www.novosti.rs/vesti/naslovna/drustvo/aktuelno.290.html:472798-Ravna-gora-i-dalje-ceka-istinu (accessed 1 October 2018).

Večernje novosti (2015) "Vučić: Slavimo Dan pobede jake, pobedničke Srbije". Available at: www.novosti.rs/vesti/naslovna/drustvo/aktuelno.290.html:547326-Vucic-Slavimo-Dan-pobede-jake-pobednicke-Srbije (accessed 11 July 2019).

Vesti online (2013) "Draža je samo branio svoju zemlju, sve ostalo je propaganda". Available at: https://vesti-online.com/Scena/Film/361336/Draza-je-samo-branio-svoju-zemlju-sve-ostalo-je-propaganda (accessed 30 September 2018).

Vučković, Z. (2015) *Sećanja iz rata*. Belgrade: Srpska književna zadruga.

6   Unearthing the past

While painting an iconostasis in a church on the mountain of Zlatibor in October 2017, speleologist and fresco painter Dimitrije Mirko Ćelić met a local who told him a story about two mass graves in the area, containing remains of local young men executed at the end of the Second World War by the Yugoslav communists because of their belonging to the Chetniks. He immediately called speleologists from Belgrade to join him in explorations of a cave nearby and, as they climbed down to the bottom of Cerova cave, he discovered dozens of human skeletons. According to Ćelić, each had a wire around wrists (Janković, 2017b). He found various objects: shoes, belts, cigarette cases, mirrors and other personal belongings, but no military symbols, because "the Chetniks had most probably taken them off in fear of revenge and hoping to survive the war" (Janković, 2017b).

After the discovery, Ćelić, a history enthusiast, committed himself to studying the postwar executions in the area through testimonies of the oldest locals and concluded that between 200 and 300 Chetniks had been tricked and executed by the Partisans there. He called Milan Stamatović, the mayor of Čajetina, who went down to the cave with six speleologists and four priests. The priests climbed down a rope to perform a memorial service for the unknown dead at the bottom of the cave. The police were informed about the findings, in the hope for a professional exhumation and identification of the remains (Janković, 2017a).

Among the objects found in the cave were glasses resembling those worn by Dragoljub Mihailović. Immediately, the news about the possibility of a discovery of Mihailović's remains broke out. The "greatest Serbian secret", hidden for 70 years, was about to be solved (Janković, 2017a). According to Stamatović, it was very possible that Mihailović had been buried at Zlatibor as certain testimonies claimed that Slobodan Penezić Krcun, who led the action of capturing Mihailović, and Miloš Minić, the prosecutor in his trial, used to come to Zlatibor in the postwar years and ask questions about the cave. As Stamatović explains, "it was important to them to find out personally, without intermediaries, what the locals knew about the secret hidden in the cave" (Janković, 2017a). He claimed that researchers had been zealously kept away from the Cerova cave. Something important had to have happened there, such as that Mihailović had been executed and thrown into the pit where his young comrades had been buried a year earlier.

If anyone involved in this discovery had checked, they would have quickly realised that it was impossible for Mihailović to have been buried there. The glasses discovered in the mass grave could not belong to him because he had them with him at the time of his capture and belonged to the Security Intelligence Agency now. Three years before the Zlatibor discovery, the *U ime naroda* exhibition exhibited the glasses for months in the Historical Museum of Serbia and the media reported on them extensively, including *Večernje novosti* that published the exclusive news about Zlatibor in 2017. The Military Museum also presented Mihailović's glasses and other personal belongings to the Serbian public in 2010 (*Blic*, 2010a). When it became clear that this could not be Mihailović, the media interest disappeared and professional exhumation and identification of remains did not follow.

Almost a decade before the controversy of "Draža's glasses", the Serbian state authorities initiated and supported attempts to solve the greatest Serbian mystery of the location of Mihailović's remains. In a similar way to the discovery at Zlatibor, their efforts were predominantly based on testimonies of a very uncertain reliability and often ended in embarrassing failures. Involving random characters turning into important actors and exhumations of an amateur nature, the state attempts have often resembled the self-initiated action of a fresco painter and a group of speleologists.

In 2009, the Serbian government established two fact-finding commissions dealing with the gravesites of the postwar period. The centrality of the Chetniks and their leader Dragoljub Mihailović in dominant memory politics and the vast importance of finding Mihailović's burial site is evident in the fact that the first commission that the state formed had only one task – to find more information about the circumstances of Mihailović's execution, with an ultimate goal of finding his remains. In addition to that, the government established The State Commission for Secret Graves of those Killed after September 1944 (*Državna komisija za tajne grobnice ubijenih posle septembra 1944*, from now on: Commission for Secret Graves or the 1944 Commission), with a broader scope of locating and marking all graves in the territory of Serbia. Both commissions involved historians and were affiliated with the Institute of Contemporary History. In 2011, during the process of rehabilitation of Mihailović, they merged and joined their efforts searching for his remains in Belgrade, without success, however.[1]

The Mihailović Commission

The State Commission for Investigation of Circumstances of the Execution of Dragoljub Mihailović (*Državna komisija za utvrđivanje okolnosti pogubljenja Dragoljuba Mihailovića*, from now on: Mihailović Commission) formed in April 2009. It consisted of state officials, representatives of archives and museums, state security and intelligence agencies and historians, including Bojan Dimitrijević and Kosta Nikolić of the Institute for Contemporary History. Only a few days after its establishment, the government, on the proposal of the Ministry of Justice, removed the confidentiality stamp from all documents associated with the execution of Dragoljub Mihailović (Nikolić and Dimitrijević, 2009: 10). The documents became available only to the Commission members, meaning only a handful of historians who have acted as mnemonic agents and professional anti-communists at the institutional level. The same case is the Commission for Secret Graves.

The work of the Mihailović Commission went through two phases: the archival research about information on the location and circumstances of Mihailović's execution and physical search for his remains. Although a few alleged discoveries happened in 2010, the first phase had practically ended by June 2009. The Commission introduced an open telephone line, but the calls they received turned out to be worthless and the search for witnesses and their hearing was also not fruitful. During the research phase, the Commission did not find any documents revealing where Mihailović's remains could be. As historian Bojan Dimitrijević eloquently summarises:

> We did research in the Military Archives, in the archives of the State Security, the Archives of Serbia and that was it. The commission had regular meetings; a public call was made so that people who knew anything about it could get in touch. The people who knew nothing at all mostly called, we did not have a single relevant witness and we did not find anything relevant related to this topic during the archival research, so we practically finished the whole research already in June 2009 and it was already then when we, at least the two of us here in the office, realised that we would not find anything special.
>
> (Dimitrijević, 2016)

The archival research revealed possible locations in Belgrade where Dragoljub Mihailović could have been shot but there are no documents or witness statements that could confirm it (Nikolić and Dimitrijević, 2009: 13–14). A general issue the Commission members encountered was the disappearance of a segment of documents related to Mihailović's case, as Yugoslav officials had often moved, hidden, destroyed and taken them while working on their memoirs and other publications. According to Nikolić and Dimitrijević, the most interesting parts of this "special archive" disappeared (2009: 20). Interestingly, the Serbian intelligence and state security services had known about this problem before the Commission's establishment. The representatives of these institutions attended the Commission meeting in May 2009, and stated there and in the media that they did not have any relevant documents about the execution of Dragoljub Mihailović in their possession, raising concerns if such documents existed at all (Vlahović, 2009). Some Commission members, such as Milan Gavrilović, the director of Oplenac memorial complex, claimed that the documents existed, but that the state security hid them (Čalija, 2010).

Sensationalist media reports followed all activities of Mihailović Commission. Commission members, particularly Slobodan Homen (DS), the state secretary in the Ministry of Justice, contributed to the media attention by announcing great discoveries of documents that would certainly lead to Mihailović's remains. For Dimitrijević, who had known from the start that there was not much there, it was the politicians who wanted to make the Commission seem more important in the public at all costs, even by creating the impression that they were close to a discovery (Dimitrijević, 2016). The first of such alleged discoveries was an archive near Belgrade, consisting of documents related to the execution of Mihailović. Homen confirmed the story:

> In these documents, there are, among other things, photographs of the execution of Draža Mihailović, a complete documentation and transcript about the implementation of the death sentence. This is completely new information because the archive has not been opened since 1946. The photographs were made to show the regime that the death sentence was carried out. The archive is outside of Belgrade, but we do not want to reveal where exactly.
>
> (Dedeić and Arsić, 2009)

The archive did not exist and no one from the Commission mentioned it again. However, just a year later, the Commission made another discovery. A delegation of three Commission members travelled to London and, even though historians of the Second World War had extensively used the British sources, they "found" thousands of pages of documents dealing with the period from 1941 to 1945 in Yugoslavia. They had them copied and delivered to Serbia. As Homen announced, a few documents mentioned Mihailović (Bilbija, 2010). The Commission discovered a document with a double confidential stamp and the same date as the date of Mihailović's death, making Homen certain that the document was crucial for solving the mystery of Mihailović's burial site. The Commission determined the date of Mihailović's death, 17 July, through a telegram of the British Embassy in Yugoslavia informing the government about the death sentence (*Blic*, 2010b). The Commission was supposed to receive the confidential document and Homen hoped to find out within a month where Mihailović had been buried. However, nothing further happened.

The first phase of the work of Mihailović Commission entails numerous issues. First, what the Commission looked for did not exist. Their hypothesis was that there must be a document describing where Mihailović had been executed and buried. Assuming that every past event must be reflected in documents, they looked for a report, a transcript or a note someone made after the execution (Jovanović and Radić, 2009: 78–79). The search continued for months after the Commission knew that no institution or archive had such documents. Moreover, it is unclear how they concluded that a British archive could have a document with information where the Yugoslav authorities had buried Mihailović. Additionally, the Commission members continuously misled the public through announcements of alleged discoveries and subsequent silence when their explorations failed. Regardless of their failure, the media attention made the image of Dragoljub Mihailović as a victim constantly present in the public. Finally, the very existence of the Commission, even if its establishment was the result of the personal interest of state actors such as Slobodan Homen, demonstrates that the Chetniks and their defeat became such a significant historical reference in the post-Milošević period that the question of location of the remains of their leader became a state affair.

The State Commission for Secret Graves

The Commission for Secret Graves formed in July 2009. Its establishment was not the result of the initiative of state institutions or demands of a victim association, but of a joint initiative of Momčilo Pavlović, the director of the Institute of Contemporary History, and *Večernje novosti* newspaper. The state commission started as a daily feuilleton *Unearthing the Truth* (*Otkopavanje istine*), publishing lists of people executed, and burial locations accompanied by emotionally charged testimonies by either victims or their descendants. The media action ended with a roundtable discussion on crimes of communism, involving the representatives of the Institute and *Novosti* and prominent intellectuals.[2]

Two state representatives attended the roundtable, Srđan Srećković, the SPO member and the Minister of Diaspora, and Slobodan Homen, the State Secretary from the Mihailović Commission. As most of the other roundtable participants, they were known for their anti-communism and had acted as prominent memory entrepreneurs, becoming crucial for bringing the Commission to the Government's agenda. Just a couple of days after the roundtable, the Commission for Secret Graves came into existence.

With the experience of the Mihailović Commission and the political climate favourable for their objectives, anti-communist historians lobbied for another commission. The imagination of the resistance of "former communists, their descendants and supporters, infiltrated in all pores of the Serbian society", additionally motivated them to make the political decision-makers interested in "crimes of liberators" (Pavlović, 2010: 10). The fact that the notion of dealing with the past in the post-Yugoslav space usually refers to the 1990 wars is another important reason for frustration. In his opening speech at the initial roundtable, Pavlović argued for the need to deal with the postwar executions because confronting the past usually refers to the 1990 wars and implies Serbian guilt by rule, based on the pattern dominant in the West – that Serbs are the biggest and only culprits of destruction of Yugoslavia, the aggressors while everyone else are victims (Pavlović, 2010: 12).

The Commission started working in November 2009 with Slobodan Marković as president. The second most important person whose media presence made him the face of the Commission was its secretary Srđan Cvetković, at the time a doctoral student examining the repression in Serbia employed at the Institute of Contemporary History. The Institute constituted a crucial segment of the Commission as a key location where almost everything except fieldwork took place. The Commission officially consisted of two boards, responsible for research and locating the burial sites, situated at the Institute, and the Board for Exhumations dealing with forensics and identifying the remains that only existed on paper. The Commission formed the Documentation Centre with the purpose of gathering and archiving documents, photographs and audio and video files received from citizens and descendants (Pavlović and Cvetković, 2012: 15).

Purpose

The Commission for Secret Graves had a task to

> investigate, find, and mark all secret graves with remains of people killed after the 1944 liberation by: locating and marking all grave sites and determining the exact number of people; by exhumation and personal identification at those locations where possible ...; and to prepare and deliver a special report on its work to the Ministry of Justice and Government.
> (Poslovnik Državne komisije za pronalaženje i obeležavanje svih tajnih grobnica u kojima se nalaze posmrtni ostaci streljanih posle oslobođenja 1944. godine, 2009)

The formal objectives of the Commission did not imply evaluation but finding facts about the numbers of people who lost their lives after 12 September 1944 in Serbia. As Pavlović outlines, they would not determine guilt or innocence, but were

> interested in and motivated by the fact that a significant number of families whose members were shot have never received any written verdict, not even a manipulated one, and a burial site is unknown for a large part of them.
> (2010: 13)

In a similar way, Cvetković emphasises that it is a simple survey of how many people were executed, not how many innocent people were killed (Cvetković, 2016). However, he also underlines that:

> But they did, when they were shooting, when OZNA did it and during the court processes later, they are all deliberately mixed like that and put together in books and lists. They did it on purpose to cause confusion, so that if anyone would try to talk about it today, the reaction would be "you are rehabilitating war criminals", because they tied them together in this way. Why is it then always in the lists, for instance, a Gestapo agent and then some industrialist, or a priest, student, I don't know.... So, they always put two, three, four persons to compromise everyone. That's a classic UDBa technique. And that UDBa technique is used by a modern *udbaš* to discriminate and discredit people who research that.
> (Cvetković, 2016)[3]

In 2011, the Commission published its results as an online database, the latest version of which encompasses 59,554 names ("Otvorena knjiga". Registar žrtava Komisije za tajne grobnice ubijenih posle 12. septembra 1944, n.d.). The only formal criteria of the Commission's investigation and database is that the Partisans killed a person after 12 September 1944. The official documents do not use any other phrases such as persecuted, missing or perished. The Commission members predominantly use the term "shot". The database, however, involves numerous issues of inconsistency, such as people who died in combat, missing persons and people who died but were not executed. The list includes the Chetnik combatants who might have lost their lives while retreating and fighting against the Partisans in Bosnia.

The official name of the Commission illuminates the centrality of the Chetnik movement and Dragoljub Mihailović for the institutional level of memory work. Even though the specific Mihailović Commission had already existed, and the 1944 Commission had a broad scope of postwar executions in general, its title links it closely to the Chetnik movement. Namely, they took 12 September 1944 as the starting date because, as Cvetković explains,

> it was then when the Partisans with the Red Army entered Serbia and liberated it from the German occupation, and this was also when the king called all members of the Royal Army in the Homeland led by Draža Mihailović to put themselves under Tito's command.
>
> (Arsenović, 2010)

The media and the Commission representatives refer to the people whose names are in the database broadly as "victims". In this way, they indirectly represent them as innocent victims of communism executed only for ideological and political reasons. Moreover, while the database allows additional information, there are no comments for people who, for instance, had been responsible for war crimes, either directly or by command responsibility. In this sense, the Commission reflects the wider tendencies of the observation of the entire Second World War and the period state socialism through the narrow lens of the immediate postwar period, ignoring the wartime activities of the persons concerned. If they occasionally acknowledge that some people faced retribution because of their wartime activities, it is framed as the minority:

> There were, understandably, also people who no court could exonerate because they had committed crimes during the occupation, but the majority are people who would not be found guilty or sentenced even by the most unjust court. Especially not sentenced to death.
>
> (Pavlović, 2010: 13)

With all its flaws, the results of the Commission for Secret Graves have a positive effect on debates and number games about the postwar executions, the numbers of Serbs executed by the Partisans and the claims that the Partisans completely eradicated the Chetnik movement. The Commission results proved all claims wrong, particularly the narrative of Serbian civilians from central parts of Serbia as the largest victims of the postwar violence. First, the number they determined is lower than 60,000, as opposed to claims that went up to 300,000. Out of that, 68.85 per cent came from Vojvodina, the majority of which are either Germans or Hungarians, while Serbia without Kosovo and Vojvodina contained less than 30 per cent of all names.

The exhaustive study by Radanović reveals that more than 7,000 people in the database were from the Yugoslav Army in the Homeland, which comprised 49.07 per cent of all names in central Serbia (2015). Considering that the Chetniks had around 25,000 members in August 1944, according to German documents, he concludes that most Chetniks from Serbia survived the war and the postwar period (Radanović, 2015: 595). Approximately half of the Chetniks joined the Partisans in 1944 and those who had not committed war crimes received amnesty. The Commission data shows that civilians did not represent the majority, as the Chetniks and other armed forces constituted more than half of the names.

As opposed to the results of their own work, Commission members insist on arguments from the time before the database existed. In a more recent journal article, Cvetković analyses the structure of the victims, arguing that most were civilians and that the primary targets were urban citizenry, critical intelligentsia, politicians, merchants, clerks, factory owners, priests and wealthier peasants (2016). This is one of the central arguments of memory entrepreneurs and revisionist historians, that the aftermath of the 1944 liberation symbolises the destruction of Serbian political, economic and intellectual elites, as, in Cvetković's words, "the sword blade was turned to the old bourgeois class and inherited social elite of prewar Serbia" (Vasiljević, 2009). According to this view, settling accounts with collaboration and crimes was only a disguise for the specific aim of destruction of Serbian elites.

Even though the database stands in contradiction to these claims, Cvetković argues that the postwar purges were of a higher intensity and different nature in the Yugoslav case in Europe. Because of meticulous planning and having collaboration only as a pretence, Yugoslavia is incomparable to any European country as more brutal, more repressive and more systematic than elsewhere (2016: 69). Going even further, he underlines that Serbia was an exceptional case within Yugoslavia because the victims were predominantly civilians, as opposed to summary executions in Slovenia.

Results

Until its dissolving in 2015, the 1944 Commission only made two exhumations. The mapping of mass graves in Serbia resulted in more than 200 locations but none of them have been officially marked. The two exhumations that the 2010 annual report calls "tests" were entirely amateur and probably illegal, done by the Commission members, volunteers and followers (Cvetković, 2010). The first initiative took place in October 2010 near Boljevac in eastern Serbia where at least 44 people from the area were shot, according to the Commission's investigation. Obtaining the necessary tools from locals, Cvetković and two volunteers, the landowner and "interested descendants of victims" started digging, without forensic experts or responsible authorities (Cvetković, 2010). They found human bones. As Nikola Baković, who volunteered in the Commission at the time, recalls:

> People in the village were talking that people had been shot there, it was also known which local people had been taken there. So, we basically came there with two–three shovels and spades, we started digging, and, first, very soon, the digging didn't even last that long, we found flagstones. Slab stones under a very thin layer of soil. However, they got a bit fragmented over the time, so they were not that bulky anymore, so three or four of us could move them away. There was more dirt underneath. Soon after that, we found certain, how I should put it, things that looked like bones. We couldn't figure out if they were bones or not at first but then we found something that looked like a skull fragment. So, those were really someone's bones. There were quite a lot of them, a lot of those bones.
>
> (Baković, 2016)

The way the Commission unearthed this tomb was from an official fact-finding commission at work. As Cvetković summarises:

> We are digging with a brittle shovel in the presence of an older man who came back from Munich after being in emigration for 50 years ... he came back, after 50 years, to dig out his father in a remote area in east Serbia, illegally, half-illegally, although we are the State Commission. Although we are the State Commission and I am its secretary there with a shovel.
>
> (Cvetković, 2016)[4]

The "exhumation" attracted public attention. Soon, the police and prosecutor's office were contacted to mark the location and the Commission went to the Office of the War Crimes Prosecutor, only to be told that the burial site was not in their jurisdiction. Finally, the court in Zaječar took over the exhumation in 2011, but the case stopped there. In Cvetković's opinion, this case shows the lack of will of state institutions to confront the past and totalitarian legacy (2016).

The only other exhumation took place near Vlasotince in southern Serbia where around 60 people had allegedly been executed after the liberation, including the wounded Yugoslav Army of the Homeland soldiers, civilians and personnel of a nearby hospital (Cvetković, 2010: 45). Witnesses and relatives informed the Commission about the location. Once they found it, the Commission members "began probing the site with the regular shovel, the method applied in the case of Potok Zmijanac near Boljevac" (Cvetković, 2010: 45). Conducted in the same informal way, the "test exhumation" also revealed stones and human remains underneath. The local police, county prosecutor and media were informed, and state institutions, forensic experts and the Board for Exhumation took over the further investigation. The 2010 Commission report suggests that the location should have been appropriately cleaned and marked so that the descendants could come close and pay their respects (Cvetković, 2010).

Except the two rather informal attempts of unearthing remains, the Commission for Secret Graves has not done anything regarding exhumation and identification of mapped gravesites. In fact, the Board for Exhumation did not exist. During my conversation with Cvetković and Marija Ilić of the Documentation Centre, they could not recall the name of the person responsible for the Board:

> That did not function in practice, it was a hoax. The Commission cannot do it without the concrete assistance from the prosecutor's offices and courts.... And since the Commission did not have a budget, or funding, or will to do something like that, it was, how should I put it ... a dead letter.
> (Cvetković and Ilić, 2016)

The Commission only mapped almost 200 locations but did not do anything else. During the first phase of their work, Pavlović and Cvetković travelled across Serbia and visited the supposed burial sites, talked to the locals and took photos. Failing at another objective they outlined, the 1944 Commission did not mark any of these gravesites. During their visits, Pavlović and Cvetković discussed the publications that would deal with the crimes of communism in each county, suggesting to local municipalities they could erect a monument or mark the burial sites in some other way. However, as discussed in the next chapter, the marked sites are usually local grassroots initiatives or anti-communist memory activism.

Regardless of the failure to complete planned goals, the feuilleton about the postwar executions that provoked the creation of the Commission and its continuous presence in the media makes it a very powerful agent of memory. The public discourses surrounding the Commission's findings and its official position as a governmental body made it all seem much more relevant than it really was. In addition to that, the results, the gathered facts and testimonies were later made public in the form of a highly visited exhibition followed by a series of public lectures, discussions and film screenings. Finally, despite its failures, the Commission's work was constantly surrounded, supported and praised by a significant number of people from Serbia and emigration, mostly descendants and relatives of the executed, who ascribed it large symbolic importance. With the involvement of external actors, the Commission became more than just a state enterprise of fact-finding, constantly oscillating between multiple levels of memory work.

Informalities

The work of the Commission for Secret Graves did not always resemble a state-sanctioned investigative body and it should be regarded as an informal network instead. First, the lack of resources and state involvement in the daily work of the Commission, entirely outsourcing it to the Institute of Contemporary History, represents a crucial factor for the way the Commission would later function. This, together with the only formal participation of ministries and other institutions, contributed to the Commission's informal and frivolous nature, the root of most problems in the Commission's results and their public representation. The symbolic power of the Commission as a mnemonic agent was not less intense but reinforced by involving actors of the individual and social levels of memory work.

Officially, the 1944 Commission involved historians, different ministry representatives, the Security Intelligence Agency, archives, Branimir Aleksandrić of the Institute of Forensic Medicine and Dobrivoje Tomić as the victims' delegate (Cvetković, 2010). There was a lack of expertise for the topic of investigation among historians, with the Commission's secretary being a doctoral student at the time of the establishment and no one else working on the postwar executions. Additionally, most state officials were only formally affiliated. The exception is Homen as a crucial agent that made both fact-finding commissions happen, with Slobodan G. Marković as another influential person in this period. Finally, the Institute of Forensic Medicine did not engage in exhumations nor did the Board for Exhumation really exist. The common denominator of the Commission's active members was anti-communism, rooted in their political convictions or, in the case of victim delegates, in personal or family experience of the postwar retributions.

In addition to the official Commission associates, dozens of people contributed to it. As Cvetković explains, these were three categories of people: the people who actively engaged in the Commission's work with or without honorarium, the employees of local and municipality archives who volunteered to send materials and the people who sent their family archives, stories and documents (Cvetković, 2016). In the beginning, the Commission did not receive enough funding from the Ministry of Justice to employ professional research associates. The personal contacts of Cvetković became crucial for animating people across Serbia:

> When the Commission had already been established, when I came there, as I knew many people who dealt with this topic because of my doctorate, I contacted all those people, I used my position to make them a team to help me … I gathered all those people to work on quantification, I divided them into counties, so the team was growing and had dozens of associates. Everything on a volunteer basis. At some point, when the enormous documentation of BIA was opened, I demanded that something should be paid to those students, they were mostly graduate students of history. So, they were paid a little bit, maybe ten to fifteen thousand … some people entered out of curiosity, some as archivists, there were different motives. It was completely spontaneous, unpaid, volunteer. Then some worked here, some came, some left. This is how we collected … now already 59 thousand

people are in the database, and then we made cooperation with the SANU commission that only researched Vojvodina.

(Cvetković, 2016)[5]

From early 2010 onwards, the Commission had volunteers based in Belgrade who retyped the OZNA lists of executed persons, handed over to the Commission by the Security Intelligence Agency archives (BIA). The main work was retyping and copying the lists in the Archives of Serbia. As Baković explains, they retyped the lists, he updated the website and they sometimes interviewed the relatives of executed people. He also took part in the first "test exhumation" near Boljevac:

> We were basically copying how those lists originally looked like. So, there was usually the name, last name, father's name, the year of birth or the date, and there was the place where the person comes from, but it was unclear whether that means where they were born or where they lived, and this is where dilemmas often occurred, how to write that … what other categories were there … profession, the date of execution, as well as the reason, which formation they belonged to … we followed those lists and it is true that those lists could but do not have to be true, but OZNA made them so, it was written by the people who were either directly involved in executions or were superiors. That gives the lists a certain credibility.
>
> (Baković, 2016)

History students or graduates like Baković represented the minority among the people working with archival documents. For them, enthusiasm about working with the previously unresearched materials turned into disappointment about the nature of the Commission's work and its public representation. Reflecting on his time in the Commission, Baković recalls that everything became non-serious later: the way the Commission's work was promoted in the public, the people it affiliated itself with, the political emigration and flirting with anti-communism and how the Commission represented itself in the media. For him, determining how many people were killed is purely a fact-finding endeavour, but the problem lies in treating everyone as victims (Baković, 2016). Danilo Šarenac, a historian of the First World War who worked for the Commission in 2011, remembers numerous practical issues, such as months-long waiting for the equipment or symbolic salaries, but also a strange mélange of people who gathered around the Commission:

> Just retyping.… It was really tiring because everything had to be checked once or twice to see if someone had maybe been left out. We had a set norm, some number of names that would be adequate, so we really worked and Zorica really kept those standards, so that literally there were no breaks.[6] There was a fine difference between historians and, let me say it, amateurs because she brought coffee to the reading room of the Military

Archives, which was strictly forbidden, but no one rebuked us, and it was interesting because no historian would do that. It made my hair stand on the end, but I didn't want to say anything.

(Šarenac, 2016)

The lists of executions were crosschecked with other materials: secondary sources, materials received from archivists and other enthusiasts and testimonies of victims, descendants and witnesses. While working on his master's and doctoral thesis, Cvetković contacted many people whose family members had faced persecution in the postwar period. When the Commission formed, they got involved, mainly through Marija Ilić who personified the Documentation Centre of Victims (*Dokumentacioni centar žrtava*) and was the only person on the regular payroll by the Ministry of Justice, although without professional expertise.

The interest of people with direct or family experience of the postwar retributions was immense. A few hundred people passed through it. Sometimes, they formed queues in front of Cvetković's office at the Institute for Contemporary History. The Documentation Centre, namely Marija Ilić, met the interested parties, took information and materials from them and organised them into folders (Cvetković and Ilić, 2016). The testimonies of descendants (*potomci*), how the Commission members refer to them, were crosschecked with archival documents and the database. A testimony often confirmed something already found in the documents and sometimes the Commission found documents confirming the testimony afterwards. Besides gathering testimonies, Ilić explained to the descendants what rights they had, informing them of the possibility of judicial rehabilitation. She considered the testimonies crucial for the Commission's work:

> Documentation is necessary, and archives are necessary, but, without the living descendants and their memories, without them, everything seems prosaic, it doesn't seem credible. They were crucial for us, not because of the information, but to help us experience that time in a different way. Historians will provide statistics, how many people were killed, how many people imprisoned ... but those are just numbers. What we all need is taking a step back in time through those memories, to see what really happened, if it was justifiable, then why was it justifiable, if it was not justifiable, why was it not. Because, the Commission did not make judgements, it was not made to judge the executors, it was also not made only to praise the victims, but to show us the time that was as it was. Many innocent people perished, many innocent people perished overnight, from 20 October mostly until April, June 1945. This story of the internal civic suffering is the hardest for us. This is why it is necessary to look at what was done back then and face it, to look at the numbers but also to hear the experiences of descendants, people whose property was taken away, who were kicked out in the street, young girls who perished just because they had a French last name, for

instance, 16 year old little girl imprisoned, legs and arms broken, raped....
A completely unjustifiable use of force.

(Ilić, 2016)

The importance of these testimonies for the Commission lies in the fact that they were the main source of information about all mapped burial sites. The OZNA documents rarely included insinuations about locations of burial sites, making oral testimonies crucial for the mapping of graves. However, the Commission (as well as Cvetković in his research) accepts the testimonies with all their details uncritically, without questioning their reliability. The texts the Commission has generated, including its official reports, very often use the vague phrases such as "it is believed", "the locals think" or "according to our estimates" when talking about the locations. The lack of exhumations further complicates this issue.

The testimonies did not only help the Commission, but they transcended it. The media, public talks and scholarly publications reproduced a few dramatic and very vivid stories of executions, imprisonments and torture. These stories became a pillar of the 2014 exhibition. Their significance is not their factual accuracy but the emotional power, demonstrating the brutality of the new regime and the innocence of the victims. Some of these stories of suffering have been repeated so much that they have become a commonplace in Cvetković's publications and talks, media and public outputs of different anti-communist groups.

The epilogue

In a similarly unusual way like its establishment, the Commission for Secret Graves ceased to exist in 2015, although it had practically stopped working in 2012. After years of constant media presence, it withered away silently. Cvetković was in the process of resignation for years. During the interview with him in March 2016, he was not aware of the fact the Commission had not existed since the end of 2015. The official statement of the Ministry of Justice claimed that the Commission did its job independently and on its own and the only obligation of the Ministry was to provide administrative support to its work, encompassing the employment of one person on the recommendation of the Commission's president (Jeremić, 2016).

The informal nature of the Commission, the involvement of a wide circle of people, Cvetković's status within this group and his personal motivation made the Commission transcend the state-sanctioned fact-finding endeavour and continue existing. First, its findings generated the previously mentioned travelling exhibition in the House of Terror style in 2014, visited by more people than any other museum project in Serbia and accompanied by numerous lectures, film screenings and discussions. The exhibition balanced on numerous levels of memory work, with the partial state support and bottom-up efforts.

Srđan Cvetković worked tirelessly on the dissemination of his work through hundreds of public lectures also entitled *U ime naroda* that he gave across whole

Serbia, in city halls, libraries, schools, in churches and offices of the SPO. His lectures gather people brought together by their self-identification as victims of communism or their anti-communism, including libertarian and monarchist organisations. Finally, the findings of the 1944 Commission and everything related to the exhibition, including its name, social network profiles and website, transformed into *In the Name of the People – For Free Serbia* association of citizens (*U ime naroda – Za slobodnu Srbiju*). The association is led by Srđan Cvetković and presents the 1944 Commission's results, such as the database, on their website and other public outlets. The association, examined in the next chapter, involves the people who gathered around the 1944 Commission.

The quest for the grave of Dragoljub Mihailović

> And then, after a few hits with the shovel by the author of this text, the spade in the hands of Blažo Đurović pulled out around eight centimetres long human bone at fourteen minutes past two o'clock, a sign to the group of diggers that they are on the verge of solving the dark secret and that it is possible, at least in this case, to neutralize the Bolshevik punishment of deprivation of a right to grave.
>
> (Četnik, 2011)

The state commissions responsible for the investigation of secret graves and the execution and burial site of Dragoljub Mihailović joined their efforts in 2011 with the goal of finding Mihailović's remains, as the process of his rehabilitation was gaining momentum. When the joint initiative of the two state bodies reached the fieldwork phase in June 2011, different individuals with vested interest in the outcome of the search got involved. Announcing the usage of "3D technology" that would enable digital section of the soil, Cvetković called Mihailović's grave one of the biggest Serbian secrets (*Večernje novosti*, 2011a). Followed by enormous media attention, speculations and sensationalist statements, the physical search for Dragoljub Mihailović ended unsuccessfully. This initiative of the state-sponsored actors reflects the informal and often amateur nature of the state commissions. It is also another example of the image of Dragoljub Mihailović as a symbol of the narrative of Serbian victimhood under communism and the central motif of contemporary memory politics.

In April 2011, the Mihailović Commission presented the results of their investigations to the public. After two years of work, the Commission could only conclude that Dragoljub Mihailović had been executed early on the morning of 17 July 1946 and had been buried in Ada Ciganlija in Belgrade, next to a prison that was later demolished. During their research in Serbian and foreign archives, the Mihailović Commission had not found any direct evidence about the time and the place of the 1946 execution, deducting their conclusions from a variety of written and oral sources. Claiming that Mihailović's remains had been moved and buried at another location, Homen announced probing of the primary location in Ada Ciganlija (Bilbija, 2011). In the opinion of Slobodan Radovanović, the republic public prosecutor who acted as the Commission president, the secondary grave could probably never be found, but he emphasised the significance of the Commission's work as correcting a great historical injustice done to Mihailović (*Radio televizija Vojvodine*, 2011).

The first probing of the location in Ada Ciganlija on 18 June 2011 resembled the "test exhumations" of the 1944 Commission. Similarly to Boljevac, Cvetković led this exhumation attempt that was amateur, without forensics or archaeology specialists, and it involved a random group of people, including journalists of *Blic* and *Politika* newspapers and passers-by, who picked up shovels and spades. In addition to that, Blažo Đurović, the vice-president of the Association of Political Prisoners and Victims of Communism, provided his company's land surveying equipment. Not affiliated to any state body or an exhumations expert, Đurović became a prominent actor of the Ada enterprise, giving media statements and opinions on forensic issues. His equipment, that Cvetković refers to as "3D technology", had already been used by the 1944 Commission a few months earlier. As Đurović explains:

> I accidentally had an opportunity to see probing and exhuming of mass graves in Slovenia by modern technology and, soon after that, I saw Srđan Cvetković unearthing a mass grave with a shovel. This is where I decided to make all my capacities available to commissions in Serbia for free.
> (Kovačević, 2011)

A few days earlier, Momčilo Pavlović announced the beginning of explorations at Ada Ciganlija as locating where the prison foundations used to be, followed by exhumation and a graduate investigation of the area at a later stage (Spasojević, 2011c). However, the group that met at Ada did not wait for a professional exhumation, they instantly started digging around the place where the prison used to be, and they found bones. Without any analysis of the findings, they all assumed that they found human bones. Media coverage started immediately, considering that journalists took part in digging up the bones. Leading into his report published in Politika in the evening of 18 June, Milan Četnik writes:

> Whether the human bones, found yesterday next to a rose bush at Ada Ciganlija in Belgrade, represent a part of the remains of Draža Mihailović, will be shown by a DNA analysis. The assumption that it is a secret tomb on that spot was confirmed by the discovery of primitive wire handcuffs that were used to tie the wrists and arms of victims of the postwar Partisan torture in the prison complex of Ada Ciganlija.
> (Četnik, 2011)

The actors involved in the discovery did not refer to their findings as anything else but human bones. The stories of wire handcuffs, of remains deliberately destroyed by communists using fire and of Mihailović being thrown in a mass grave with other highly prominent Serbs spread quickly. The Commission did not deny them but rather contributed to the belief that Mihailović had been buried there. Changing his theory of primary and secondary burial sites, Cvetković now claimed there was a possibility that it was Mihailović. He argued that the discovery of human remains at the location where Mihailović and his comrades had been executed opened the possibility that there had been no moving at all, but it was a "scam to hide the real truth" instead (Miladinović, 2011). Because of the burned bones and pieces of glass found in the soil, Cvetković explained that the Molotov cocktail had burned the bones (Četnik, 2011).

Without waiting for detailed investigation, the Commission members and affiliates acted in public as if they knew everything about their discovery at Ada, going into detail in their description. They presented mere speculations as facts. In this way, Blažo Đurović described that more than ten bones were found, human arms and hands, with metal and wire as the handcuffs with which the victims had been tied (*Danas*, 2011). In another newspaper, he claimed that the first discovered remains had belonged to older people (Miladinović, 2011). Slobodan Homen seemed to have even more information:

> It is a mass grave of a three times four meters perimeter where seven to nine persons were buried together with handcuffs. This is a proof that they were prisoners who were most probably also executed at that place, and the remains of quicklime and burnt bones point out to the attempts to destroy traces. There are remains of numerous skeletons and the DNA analysis will show who they exactly are. This case now completely goes under the jurisdiction of the Public Prosecutor's Office that will continue to cooperate successfully with the State Commission. It is important to precisely determine whom the remains belong to because, according to the witness testimonies, the highest representatives of the previous government and generals had ended up in that mass grave. Their families deserve to know the truth.
>
> (Spasojević, 2011a)

Serbian media published vivid stories daily of Mihailović's victimisation, such as that his body had been cut in pieces by axe, while the bodies of others executed in the same night had remained intact (Spasojević, 2011b). The focus on Mihailović's murder by communists completely sidelined the context of the Second World War. The question of the location of his burial was narrated as the biggest mystery, finally solved. These discourses portrayed the Yugoslav authorities, referred to simply as "the communists", as perpetrators who not only deprived Mihailović of his life but also of a right to a proper burial and a grave.

Before professional examination of the Ada findings, the actors of the "test exhumation" financed a trial forensic analysis on their own that allegedly proved that the bones were human (Spasojević, 2011b). Nevertheless, the state institutions, including the prosecutor and police, had to get involved in the process leading up to the professional exhumation and examination of the findings at the Institute for Forensic Medicine in Belgrade. As the Ada team and state commissions were primarily interested in the question of whether it was Dragoljub Mihailović, they reached out to his grandson Vojislav Mihailović for a DNA sample. Mihailović, who had been critical of the Commission before and called it "a yet another unserious attempt of the state", refused it (Roknić, 2009). The idea to exhume Dragoljub Mihailović's parents appeared as a backup plan.

Exhuming Mihailović's parents and a DNA analysis were finally unnecessary because the analysis of the bone fragments showed that they were not human. It took only a day or two for the Anthropology Lab of the Faculty of Medicine in Belgrade to conclude that the bones could not belong to Dragoljub Mihailović because they were animal bones (*Večernje novosti*, 2011b). While other members of the test exhumation team, previously prominent in media, were silent after this discovery, Cvetković spoke out, explaining that their task had never been to find the remains but only to check whether the results of the Mihailović Commission had been correct (*Večernje novosti*, 2011b).

Already before the self-initiated digging initiative at Ada, the High Court in Belgrade instructed the Institute of Forensic Medicine to conduct examinations at Ada Ciganlija. A professional team composed of anthropologists, archaeologists and other forensic experts was established. This was highly unusual for the Serbian context, as such exhumation teams had not been formed many times, the first and most prominent case being the excavations of mass graves of the war in Kosovo in Batajnica in 2001. They started working in Ada Ciganlija in August. The area of examination, a large space around the former prison walls, looked entirely different than in June 2011, fenced and guarded by the police, divided into squares and with the archaeological equipment and tools, a ground-penetrating radar (from Blažo Đurović's company) and computers lying around. No passers-by could enter the site. Instead of the expected excavation of a body, the multidisciplinary team engaged in a forensic search for an unknown grave (Bubnjević, 2011).

The expert excavations completely negated the main conclusions of the Mihailović Commission's two-year work in just one month. The careful examination of the area around the prison revealed that Mihailović had never been buried there. In addition to that, the Commission's first theory of Ada as the site of execution and primary burial site and another location as a secondary gravesite where Mihailović had been moved also proved implausible. Namely, the existence of a gravesite at Ada, regardless if moved later or not, would have left traces visible to researchers. The lime pit had probably been only occasionally used for producing lime plaster for painting the guard's house and was too shallow to be used for anything else (Bubnjević, 2011). Another six pits were discovered, however, of a more recent date (Djurić and Starović, 2015: 156). All discovered bones belonged to animals. As Marija Đurić, who led the team, and Andrej Starović explain, after the field research was completed, they used Blažović's ground-penetrating radar to scan the entire area and then excavated all places where disturbances were noticed. All irregularities were, however, caused by the city water supply works (2015: 156).

Informalities and failures of official fact-finding endeavours

The two state-sanctioned investigative commissions entail numerous mutual characteristics. They both proved unsuccessful in fulfilling their purposes. The 1944 Commission did not mark or exhume any of the burial sites it had mapped. Its final database, lacking data for the biggest cities in Serbia, contains many inconsistencies and names are mixed together regardless if they belong to civilians, members of the Chetniks or other armed or police forces, if executed by the Partisans, gone missing or killed in combat. The work of the Mihailović Commission validated certain previously known details about the arrest and execution of Dragoljub Mihailović but did not discover the site of his execution or burial. Every alleged archival discovery publicly promoted with optimism ended in vain. Finally, the joint initiative of the two commissions in the search for Mihailović's remains at Ada Ciganlija was an embarrassing public debacle after the discovery referred to as Mihailović's remains turned out to be animal bones.

The state commissions illuminate the centrality of the Chetnik movement within the politics of memory in post-Milošević Serbia as well as the superficial nature of the state interest for all other groups considered victims of communism. The entire Chetnik movement is encapsulated in the image of Dragoljub Mihailović, who takes the highest position in the hierarchy of victims of communism and is, thus, entitled to a commission dedicated specifically to the search for his grave. The fact-finding commission dedicated to all victims of the postwar retribution in Serbia was established only afterwards and not because of state efforts but lobbying by historians and a newspaper. The lack of resources and state involvement that followed contributed to the informal, frivolous, and personal dimension of the 1944 Commission.

The commissions had a dual nature of the official and informal and of the institutional and social level of memory work, oscillating between a governmental body and a bottom-up initiative. The state involvement in the regular commissions' work came down to Slobodan Homen, whose personal dedication was crucial for their establishment and funding, while other state actors participated only formally. A small group of historians of the Institute of Contemporary History emerged as central actors in both commissions. Numerous people aided the official members of the commissions, lacking necessary expertise, but with political and personal motivation and interest in the outcome of the official fact-finding endeavours. The congruity of the commissions came from strong anti-communism that all actors involved had in common.

Even though they represented strong mnemonic agents with significant symbolic power, the state-supported efforts in the sphere of fact-finding and their outcomes are a failure. They did not lead to marked graves, memorials or the location of the grave of Dragoljub Mihailović. Different actors who got involved for various reasons received nothing in return, the state failure eventually convincing them to organise themselves as memory entrepreneurs at the social level of memory work. The following chapter is dedicated to these non-state initiatives.

Notes

1 Unlike the Serbian case, the fact-finding initiatives of gravesites from the more distant past in other countries have excavations as the central task, rather than only a formal objective. The examples of successful initiatives exist in the post-Yugoslav space, with exhumations of the mass graves of the postwar period in Slovenia as the most prominent example. Numerous burial sites have been exhumed in Croatia too. While the Serbian commissions looked up to the Slovene case, the seriousness and results of the investigative work in these two countries are incomparable.
2 Čedomir Antić, Slobodan Antonić, Vladimir Cvetković and Radoš Ljušić.
3 Udbaš usually refers to a person working or having worked for the UDBA, the Yugoslav state security, as well as the state security in Serbia. It is also very broadly for people from the state security and police sector. It has a pejorative meaning.
4 The man Cvetković refers to here is Dobrivoje Tomić, also a member of the Commission as a representative of victims.
5 The 1944 Commission included 27,000 names results from the Vojvodinian Survey Committee in their database. When the bilateral commission was formed, Cvetković became its coordinator for the Serbian side and used the overlaps in the two commissions and the Vojvodina resources for the 1944 Commission.
6 Zorica Marinković was invited to join the Commission by Cvetković and she worked with him on research and gathering testimonies. Her motivation comes from the fact that her grandfather was killed in 1944. Gaining always more prominence, she was listed as a project coordinator for the *U ime naroda* exhibition in 2014 and made a film and a variety of YouTube videos for it. In 2017, she released a documentary on Goli Otok.

Bibliography

Arsenović, J. (2010) "Ozna streljala, udbaš otkrio spisak žrtava". Available at: https://vesti-online.com/Vesti/Srbija/75425/Ozna-streljala-udbas-otkrio-spisak-zrtava (accessed 23 August 2018).
Baković, N. (2016) Interview by author.
Bilbija, B. (2010) "Tajni dosije o Dražinoj smrti otkriven u Londonu". Available at: www.politika.rs/sr/clanak/139599/Drustvo/Tajni-dosije-o-Drazinoj-smrti-otkriven-u-Londonu (accessed 24 August 2018).
Bilbija, B. (2011) "Dražu, i zvanično, traže na Adi Ciganliji". Available at: www.politika.rs/scc/clanak/174049/Друштво/Дражу-и-званично-траже-на-Ади-Циганлији (accessed 3 September 2018).
Blic (2010a) "Dražine lične stvari prvi put na izložbi u Vojnom muzeju". Available at: www.blic.rs/kultura/vesti/drazine-licne-stvari-prvi-put-na-izlozbi-u-vojnom-muzeju/qfl65vp (accessed 7 September 2018).
Blic (2010b) "Srbija čeka dokument o mestu streljanja Draže Mihailovića". Available at: www.blic.rs/vesti/drustvo/srbija-ceka-dokument-o-mestu-streljanja-draze-mihailovica/47xkwgs (accessed 3 September 2018).
Bubnjević, S. (2011) "Tri kutije i sedamnaest čaura". Available at: www.vreme.com/cms/view.php?id=1010677 (accessed 6 September 2018).
Čalija, J. (2010) "'Dražina' komisija u rasulu". Available at: www.politika.rs/sr/clanak/136215/Drazina-komisija-u-rasulu (accessed 3 September 2018).
Četnik, M. (2011) "Tajna grobnica na Adi Ciganliji". Available at: www.politika.rs/sr/clanak/181386/Tajna-grobnica-na-Adi-Ciganliji (accessed 3 September 2018).
Cvetković, S. (2010) *Godišnji izveštaj 2010*. Belgrade: Državna komisija za tajne grobnice ubijenih posle 12. septembra 1944., Institut za savremenu istoriju.

Cvetković, S. (2016) Interview by author.
Cvetković, S. (2016) "The 'Savage Purges' in Serbia in 1944–1945, with a Brief Consideration of Yugoslavia as a Whole", *Review of Croatian history* XII(1): 67–102.
Cvetković, S. and Ilić, M. (2016) Interview by author.
Danas (2011) "Na Adi nađene kosti, metal i kreč". Available at: www.danas.rs/drustvo/na-adi-nadjene-kosti-metal-i-krec/ (accessed 4 September 2018).
Dedeić, S. and Arsić, V. (2009) "Grobnice svuda po Srbiji". Available at: www.pressonline.rs/info/politika/67957/grobnice-svuda-po-srbiji.html (accessed 24 August 2018).
Dimitrijević, B. (2016) Interview by author.
Djurić, M. and Starović, A. (2015) "Forensic Archaeology in Serbia: From Exhumation to Excavation". In Groen, M., Márquez-Grant, N. and Janaway, R. (eds), *Forensic Archaeology: A Global Perspective*. Hoboken: Wiley-Blackwell, pp. 149–159.
Ilić, M. (2016) Interview by author.
Janković, N. (2017a) "Naočare đenerala Draže pronađene u jami Cerova?" Available at: www.novosti.rs/vesti/naslovna/reportaze/aktuelno.293.html:694670-Naocare-djenerala-Draze-pronadjene-u-jami-Cerova (accessed 6 September 2018).
Janković, N. (2017b) "Predali se partizanima, pa završili na dnu jame!" Available at: www.novosti.rs/vesti/naslovna/reportaze/aktuelno.293.html:689418-Predali-se-partizanima-pa-zavrsili-na-dnu-jame (accessed 6 September 2018).
Jeremić, V. (2016) "Pronađene sve tajne grobnice". Available at: www.danas.rs/drustvo/pronadjene-sve-tajne-grobnice/ (accessed 21 August 2018).
Jovanović, M. and Radić, R. (2009) *Kriza istorije – srpska istoriografija i društveni izazovi kraja 20. i početka 21. veka*. Belgrade: Udruženje za društvenu istoriju.
Kovačević, M. T. (2011) "Traže masovne grobnice". Available at: http://91.222.5.72/vesti/naslovna/drustvo/aktuelno.290.html:327713-Traze-masovne-grobnice (accessed 4 September 2018).
Miladinović, V. (2011) "Rešena misterija posle 65 godina: Draža iskopan na Adi". Available at: www.pressonline.rs/info/politika/165472/resena-misterija-posle-65-godina-draza-iskopan-na-adi.html (accessed 4 September 2018).
Nikolić, K. and Dimitrijević, B. (2009) "Zarobljavanje i streljanje generala Dragoljuba Mihailovića 1946. godine – Nova saznanja o arhivskoj građi", *Istorija 20. veka* (2): 9–21.
"'Otvorena knjiga'. Registar žrtava Komisije za tajne grobnice ubijenih posle 12. septembra 1944. (n.d.)". Available at: www.otvorenaknjiga.komisija1944.mpravde.gov.rs/ (accessed 21 August 2018).
Pavlović, M. (2010) "'Zločini oslobodioca' – Zadatak srpske istoriografije visokog prioriteta", *Istorija 20. veka* (3): 9–24.
Pavlović, M. and Cvetković, S. (2012) "Istraživanja državne komisije za tajne grobnice ubijenih posle 12. septembra 1944", *Istorija 20. veka* (3): 9–17.
Poslovnik Državne komisije za pronalaženje i obeležavanje svih tajnih grobnica u kojima se nalaze posmrtni ostaci streljanih posle oslobođenja 1944. godine (2009). Available at: www.komisija1944.mpravde.gov.rs/cr/news/vesti/poslovnik-komisije.html (accessed 20 August 2018).
Radanović, M. (2015) *Kazna i zločin. Snage kolaboracije u Srbiji*. Belgrade: Rosa Luxemburg Stiftung.
Radio televizija Vojvodine (2011) "Primarna grobnica generala Mihailovića na Adi Ciganliji". Available at: www.rtv.rs/sk/drustvo/primarna-grobnica-generala-mihailovica-na-adi-ciganliji_248862.html (accessed 3 September 2018).

Roknić, A. (2009) "Pronađeni živi svedoci likvidacije Draže Mihailovića". Available at: www.danas.rs/hronika/pronadjeni-zivi-svedoci-likvidacije-draze-mihailovica/ (accessed 3 September 2018).
Šarenac, D. (2016) Interview by author.
Spasojević, V. C. (2011a) "DNK vodi do Draže Mihailovića". Available at: www.novosti.rs/vesti/naslovna/drustvo/aktuelno.290.html:334952-DNK-vodi-do-Draze-Mihailovica (accessed 4 September 2018).
Spasojević, V. C. (2011b) "Dražu su isekli sekirom?" Available at: www.novosti.rs/vesti/naslovna/aktuelno.69.html:335456-Drazu-su-isekli-sekirom (accessed 4 September 2018).
Spasojević, V. C. (2011c) "Potraga za Dražinim ostacima". Available at: www.novosti.rs/vesti/naslovna/drustvo/aktuelno.290.html:334782-Potraga-za-Drazinim-ostacima (accessed 4 September 2018).
Vasiljević, P. (2009) "Ubijani noću, u potaji". Available at: www.novosti.rs/vesti/naslovna/aktuelno.69.html:242369-Ubijani-nocu-u-potaji (accessed 21 August 2018).
Večernje novosti (2011a) "Dražin grob nikad neće naći". Available at: www.novosti.rs/vesti/naslovna/drustvo/aktuelno.290.html:327110-Дражин-гроб-никад-неће-наћи (accessed 3 September 2018).
Večernje novosti (2011b) "Iskopane kosti nisu Dražine". Available at: www.novosti.rs/vesti/naslovna/drustvo/aktuelno.290.html:338879-Iskopane-kosti-nisu-Drazine (accessed 5 September 2018).
Vlahović, D. (2009) "Tajne službe ne znaju gde je Dražin grob". Available at: www.politika.rs/sr/clanak/87924/Drustvo/Tajne-sluzbe-ne-znaju-gde-je-Drazin-grob (accessed 3 September 2018).

7 Anti-communist memory politics from below

> I waited in vain that the democratic authorities correct the injustice of the communist system and at least acknowledge me and other dissidents for contributing to its downfall. Nothing happened. The old communists took over the new parties and became their leaders. Instead of one, we have gotten numerous communist parties, non-democratic, autocratic....
>
> (Ivan Ivanović in Lovrić, 2013)

In front of a monastery in eastern Bosnia and Herzegovina, approximately 20 kilometres from Višegrad, there is a memorial to Dragoljub Mihailović. Made of concrete and painted in gold, the statue is in poor condition. Besides the low quality of the materials, the damages to the memorial might also be the consequences of its tumultuous life. The statue was built in Belgrade in 1991 and taken to the Croatian city of Vukovar where it was placed at the entrance to the city in 1992. It remained there until the end of the war in Croatia. In 1998, the statue moved to the Bosnian town of Brčko, as the Serb administration, facing the return of non-Serb refugees, wanted to symbolically secure the town by erecting the statue at a central location (Bieber, 2006: 139).

However, the golden Mihailović statue could not stay in Brčko either. According to the Law on Monuments and Symbols of Brčko District, passed in September 2003, the monuments without the support of all three constituent nations could not remain on publicly owned land. Thus, the local Ravna Gora Movement had to remove the monument in December 2003 (Matijević, 2003). Originally planned for the cemetery owned by the Serbian Orthodox Church, Ravna Gora Movement of Brčko decided to send the statue to their Višegrad counterparts instead. The statue was placed near the location where Mihailović had been captured in March 1946, while the Serbian cemetery in Brčko got another monument of Mihailović that still stands there today, not representing a problem because it is not a publicly owned land.

The revelation ceremony of the golden statue was on 17 July 2004, the anniversary of Mihailović's death. The Ravna Gora Movement from Serbia and Republika Srpska, mostly veterans of the 1990s wars, attended together with clergy led by Metropolitan Dabrobosnian Nikolaj, who gave his blessing. In the meantime, the surroundings grew into a memorial complex, including a souvenir shop and St. Nicholas Monastery. Every summer, hundreds of Ravna Gora Movement members in black uniforms with Chetnik symbols gather to commemorate the death of Dragoljub Mihailović. Here, they celebrate the Chetniks as a Serbian, not a Yugoslav, army, with the emphasis on Christianity as their important characteristic (*Hram svetog oca Nikolaja – Draževina*, n.d.).

The golden statue of Mihailović, with its fascinating moving history, is just one example of numerous Chetnik memorials initiated by non-state actors. While Ravna Gora represents the most renowned case in Serbia, there are dozens of other memorials to the Chetniks and people executed by the Partisans in the aftermath of the Second World War. All of them emerged from the individual, small-group or social level of memory work and, even though they sometimes enjoy support of the local municipality, they are not a result of state initiative. This level of memory work is where commemorations take place, of both victims of communism and defeated Second World War forces that are usually intertwined. This is also the level of memory work where erecting or lobbying for memorials to these actors takes place.

Despite the occasional support of the state or local municipalities, usually by the SPO, commemorative practices of the defeated and victims of communism remain in the sphere of vernacular culture. This sphere involves both personally and politically motivated actors. They are not homogeneous regarding their politics nor concerning the interpretations of historical events. The factor that unites them is anti-communism, which is why I refer to their activities as the anti-communist memory work.

The lack of commemorations dedicated to victims of communism at the institutional level and the symbolic nature of the existing mechanisms of dealing with the communist past has pushed them to take the matter of commemorations into their own hands. They organise themselves into memory communities. While they operate predominantly at the small-group and social level, they aspire to climb up the mnemonic hierarchy. Because of the informal nature of state-sanctioned memory politics, overlaps between the levels of memory work happen, with the actors from below becoming active in the institutional memory work, such as the 1944 Commission whose work successfully mobilised them. As a further matter, while the sphere of anti-communist memory work gathers the individuals and groups who are not in power positions, it also involves the actors with symbolic power and authority, the Serbian Orthodox Church and the Karađorđević family.

Symbolic nature of state efforts

The post-2000 period in Serbia has been marked by the hegemony of anti-communist discourses that focus on the victimhood under communism and rehabilitation of the political and military forces opposed to the Partisans during the Second World War. The state actors have dedicated serious efforts aimed at delegitimisation of socialist Yugoslavia and the revision of the Second World War, including addressing those considered victims of communism through the possibility of rehabilitation, return of confiscated or nationalised property and fact-finding commissions. However, no official commemorations or memorials dedicated specifically to victims of communism have followed. The exception are changes in street names but although thousands of Partisan street names were renamed, it happens only rarely that they are replaced with references to their defeated wartime enemies.

There are no official holidays commemorating the victims of postwar executions and trials or the armed forces such as the Chetniks. It was only in 2005 that the state institutions officially supported the Ravna Gora gathering but it was because of the SPO, who controlled the Ministry of Culture and Foreign Affairs at the time. The changes of the official calendar removed the most important dates of the People's Liberation War, without replacing them with new holidays in relation to the Second World War and the postwar period. Serbian institutions do not officially commemorate the international remembrance days, such as the European Day of Remembrance for Victims of Stalinism and Nazism on 23 August.

In addition to the lack of commemorations and memorials, the revision of the Second World War and state socialism does not go far beyond a discursive and symbolic level. The Chetniks, represented collectively in the image of Dragoljub Mihailović as a victim of communism, constitute the central theme of dominant memory politics. Even when the narrative centres on the suffering of ordinary people and involves concrete measures of compensation, it is the political and symbolic rehabilitation of the Chetnik movement that matters the most. This issue is evident in the work of the fact-finding commissions, one dedicated solely to investigating Mihailović's death and burial, as discussed in the previous chapter, that did not lead to marked graves, memorials or the remains of Dragoljub Mihailović. Expecting recognition and satisfaction of their demands, different actors joined the commissions, only to be disappointed by the results. The failure of the state-supported initiatives in fulfilling their expectations contributed to their organisation as memory entrepreneurs. The public recognition and acknowledgement by state institutions remain the most relevant objective of the anti-communist memory work from below.

Non-state actors

In November 2017, a commemoration took place in Lisičji Potok, Belgrade, where a mass grave of the people executed in 1944 is believed to be. A cross was placed as a temporary memorial, with the inscription "To the Innocent Victims of the 1944 Revolutionary Terror". Organised by the *U ime naroda* association, the commemoration involved a religious memorial service, a choir and speeches by Matija Bećković, Dušan Kovačević and Jelisaveta Karađorđević.[1] Historians Srđan Cvetković and Bojan Dimitrijević publicly promoted and attended the commemoration.[2] The Association of Political Prisoners and Victims of the Communist Regime (from now on: Association of Victims), now a group intertwined with *U ime naroda*, was a key agent, including the descendants of the people executed in 1944. The Serbian League, a non-governmental organisation dedicated to total restitution of property, had promoted the commemoration with a series of high-quality videos involving prominent persons talking about the crimes of communism and the need to commemorate them. The members of the SPO, the POKS and the Kingdom of Serbia association endorsed the commemoration and the memorial initiative. No state or city officials were involved.

The actors who gathered at Lisičji Potok embody the anti-communist memory community that emerged in the late 1980s and early 1990s: associations of victims and descendants of those executed, anti-communist NGOs, historians and other intellectuals, monarchist groups and political parties, with the official support of the Serbian Orthodox Church and the Karađorđević family. In addition to them, the groups associated specifically with the Chetnik movement, such as the Ravna Gora Movement, engage in regular commemorative practices. During the 1990s, they were opposed to Slobodan Milošević and, similarly to the post-2000 political elites, saw his overthrow as the long-awaited end of the communist rule in Serbia. Not satisfied with the state efforts of dealing with the communist past, judging them as insufficient, these groups create their own commemorative spaces, continuously seeking recognition from the state.

Choreographers and consumers of anti-communist memory work come from across the political spectrum, with anti-communism and the conviction of failed overcoming of the legacies of Yugoslav state socialism bringing them together. Their motivation is both personal and political. The personal motivation is that many of them are either descendants of the people who faced the postwar retribution or were themselves persecuted during state socialism. Seeking recognition, justice and restitution is, thus, a personal matter for many, constituting the main reason for activism. The people who did not directly face repression, such as Slobodan Đurić, consider themselves victims of communism because communism deprived them of their family members or property:

> I am also, concretely, a victim of the communist regime because my father perished in late October, early November 1944. That means I am a victim of that regime. If my father had stayed alive, my destiny could have been completely different.
>
> (Đurić, 2016)

The political dimension involves anti-communism as the driving force behind commemorative practices, often intertwined with the personal. Anti-communism also represents the main motivation for people with no personal connection to repression under communism, for whom this memory work is a purely political matter. While the political is crucial in the anti-communist memory work, its fundamental characteristic is also that the actors are not in power positions. If they are politically active beyond the memory work, it is in the political parties that are marginal or have become marginal in Serbian politics. Nevertheless, prominent public personalities, the Serbian Orthodox Church and the Karađorđević family provide this memory community with the symbolic power.

The Lisičji Potok commemoration reflects the hegemonic discourses that look at the Second World War and decades of socialist Yugoslavia through the prism of postwar repression. In the words of Savo B. Jović, the patriarch representative and the secretary of the Synod of the Serbian Orthodox Church, the communists

> arrested without reason, sentenced without guilt and murdered with no mercy, proclaiming that period as the best progress in the history of our people, claiming that the life was free, beautiful, and good then, which some, unfortunately, repeat even today.
>
> (*Mondo Portal*, 2017)

Another common reference is Goli Otok island, portrayed as a site of suffering under communism without specifying that it was a prison for the Communist Party members accused of siding with the Soviet Union in the 1948 split. Appropriated by political elites and memory choreographers from below across the post-Yugoslav space since the 1980s, Goli Otok became a tool to reduce the complex issue of the 1948 split and its consequences to the narrative of national victimisation. The association of former inmates in Serbia is not active within anti-communist memory communities, except individual members such as author Dragoslav Mihailović who opened the *U ime naroda* exhibition in 2014 (Nježić, 2014). While Srđan Cvetković's work, the exhibition and a documentary made by his associate Zorica Marinković, all engage with Goli Otok, the Association of Victims refuses to cooperate with the association of former inmates of Goli Otok, seeing them as communists and because a large number of them had worked for OZNA before 1948 (Đurić, 2016).

The anti-communist memory work from below mirrors the discourses central to state-sanctioned memory politics after 2000, with roots going back to the late 1980s and early 1990s. These discourses involve a negative reinterpretation of socialist Yugoslavia as a dark episode of national history and the revision of the Second World War as its main legitimacy source. Memorialisation of the postwar trials and executions involves the defeated armed and political forces of the Second World War, now remembered as innocent victims of revolutionary terror. Nevertheless, although the anti-communist memory community promotes the historical interpretation that resembles state-sanctioned politics of memory, they remain in the sphere of vernacular culture and see themselves as the resistance to the persisting hegemony of the Yugoslav memory culture and politics.

Communism that will not go away

One of the effects of the discontent with the state efforts after the fall of Milošević is the idea of the continuous hegemony of the Yugoslav memory culture. The anti-communist memory community considers the lack of official holidays, commemorations and memorials as a particularly important issue. Perceiving Slobodan Milošević as the last communist ruler of Serbia, his overthrow was supposed to represent the final defeat of communism. In parallel to this understanding, communism is seen as still present in Serbian society. This idea arose from the expectations that the year 2000 would bring changes, particularly regarding the acknowledgement of victimhood under communism and condemnation of the Yugoslav regime and, thus, relates to the disappointment with state efforts. The year 2000 signified the apparent fall of communism as it was believed the system had not been defeated. In this way, Gojko Lazarev, a judge from Šabac and one of the initiators of the local memorial church for victims of communism, argues that communists are still in power, as well as their children and grandchildren:

- Where is that communism today?
- In the system. In the Government. In the Parliament. Everywhere. Everywhere where there is no rule of law.

(Lazarev, 2016)

As noted above, small-group and social levels of memory activism hold the conviction that communism has not been defeated. These actors engage with the communist past "not to confirm their mastery over a now demonised and banished political system, but in order to confront and weaken its continuing hold on present", or its imagined presence (Mark, 2010: xiii–iv). Within the narrative of the whole Serbian society suffering from the trauma of the civil war that had started in 1941, communism can be both the cause of collective trauma from the past and the object of fear in the present. According to this view, the SPS, Partisans' descendants and commemorations honouring the People's Liberation War embody the firm grip of communism on Serbian society. The proponents of this narrative blame the imagined dominance of communism for obstructing the process of overcoming the past.

The memory choreographers of this sphere, when asked to specify what they mean by "the communists in power", gave a twofold answer. First, they blame the SPS as the formal successor of the League of Communists of Yugoslavia in Serbia for undermining their initiatives. For instance, for the Association of Victims, it is the fault of the communists in the City Council of Belgrade that their memorial proposals were rejected (Đurić, 2016). Additionally, the notion of inherited communism represents the claim that descendants of the Partisans or Yugoslav officials inherited not only the privileges of their ancestors, but also their ideology. According to this view, the communists' descendants in power positions prevent overcoming the past because they want to keep the privileges and property they earned during communism. Having ancestors who fought in the People's Liberation War can easily lead to such discreditation. Finally, commemorative practices from below that celebrate the People's Liberation War, the existence of different forms of Yugonostalgia and the reestablishment of the holidays, such as the Day of the Liberation of Belgrade, further contribute to the imagination of communism as a dominant political force in contemporary Serbia.

Narrating Yugoslavia

Doing the anti-communist memory work based on the negative image of socialist Yugoslavia, the actors from below, similarly to political elites, tend to remain silent about their own lives during state socialism. Most of them grew up, received their education, started careers and even earned their pensions in socialist Yugoslavia. Such life stories are unsuitable to the dominant narrative of Yugoslav socialism as a long, dark night and "a dungeon of the nation". For this reason, parts of biographies are either carefully and deliberately avoided, reframed and justified to fit the larger narrative, or are narrated in broader discourses of victimhood, repression, dissidence and marginalisation and stigmatisation for belonging to "a people's enemy family".

Most of the interviewees in this research, active in memory politics as descendants, historians or otherwise, are highly educated, the majority having received their education and started their careers in socialist Yugoslavia. Most of them, however, come from urban areas and reside in Belgrade. According to judge Lazarev, the people executed in 1944 represented the best part of the Serbian society. He claims that, even though the communists had destroyed them and made themselves the new elites, the descendants never gave up. He goes on to say that "they did not surrender during communism, they turned to education and learning, many of them rose again from the ashes, got educated and acquired property and name again" (Lazarev, 2016). It was the persistency to succeed that helped them survive under communism. However, the stigma of being the descendant of a people's enemy was always there. As Cvetković explains, that was the reason OZNA had created the detailed lists of executions, to keep information and use it if necessary:

> It was the time when even walls had ears. You could not become a political leader or achieve access in philosophy or, I don't know, in social sciences that are sensitive, in history. But, you could be technical intelligentsia, maybe a director of a company, but someone could always pull out the fact that you are a child or a grandchild of a people's enemy who was shot and it could enter your biography and, if you ruffle some feathers, it could take you down.
> (Cvetković, 2016)

According to this view on Yugoslavia, the truth about their ancestors, about the Serbian elites and about the Second World War was forbidden. As demonstrated earlier in this book, the dynamic nature of the Yugoslav memory culture enabled diverse appropriations, including commemorations of those who had fallen in the Second World War as Partisans' enemies. People mourned their family members. Nevertheless, the anti-communist discourse emphasises that memory work that did not fit the Partisan myth was forbidden. Thus, stories of the leftovers of candles found at the 1944 burial sites emphasise that people knew about the gravesites and came to commemorate their family members, but authorities forced them away and made sure to keep them away by building parking lots, stadiums and other objects there. For Vuk Drašković, who had been a League of Communists member and the chief of staff for the Yugoslav president Mika Špiljak, it had been impossible to think differently in Yugoslavia:

> I could not hear the words General Mihailović had spoken before the Partisan court in Belgrade in July 1946 because I was not born yet. I grew up in the country of the Communist Party, of its leader and their truth, the only available and allowed truth. Draža Mihailović was a Serbian degenerate, war criminal, collaborator. This is all I heard from my teacher, read in books, saw in movies. I believed it, I spoke and wrote it, until the spark of hidden knowledge flew into the darkness of my trained thinking.
>
> (*Telegraf*, 2013)

Commemorative practices from below

In May 2018, photos of uniformed men of the Serbian army laying flowers at the Ravna Gora monument of Dragoljub Mihailović appeared on social networks. It caused enormous public attention in Serbia and beyond, but the Ministry of Defense issued the official statement assuring they did not officially send the army delegation to Ravna Gora and that 200 soldiers had done it on their own initiative. The delegation of four buses of the Army of Serbia must have received approval from higher ranks in the General Staff for such an initiative, but the affair remains unsolved.

Even though state authorities have occasionally supported Chetnik commemorations, such as when Oliver Antić, the advisor to Tomislav Nikolić and lawyer in Dragoljub Mihailović's rehabilitation process, advocated incorporation of Chetnik flags alongside the Partisans' in the Liberation Day parade (Živanović and Jeremić, 2014), official memory politics normally does not involve their direct and official commemoration. Numerous commemorations and memorials dedicated to the Chetniks across Serbia mostly come from below. The actors in this sphere are not in positions of power. The struggle against the perceived hegemony of the Yugoslav memory culture motivates different groups to develop their own commemorative spaces while fighting for the public and official recognition.

The struggle against the Yugoslav mnemonic hegemony does not involve the physical destruction of Partisan monuments, which in Serbia happens as an individual act directed against the Yugoslav memory culture, rather than as a tendency. For instance, Miloslav Samardžić, from *Pogledi* newspaper and publishing house, smashed dozens of red stars with a pickaxe in the Šumarice Memorial Park in Kragujevac (2016). More often than deliberately destroyed, the Partisan memorials in Serbia are simply neglected. While memorials in some cities moved from central to more marginal locations, such as the statue of Tito in Užice removed from the main square in 1991, the mass removal and destruction of communist statues that has transpired across the post-Soviet space is not the case in Serbia.

In Croatia, however, the removal of red stars from local memorials dedicated to victims of fascism is more common. These actions were not the result of staunch anti-communism but were related to the change in meaning of the red star, now worn by the Serbian Yugoslav People's Army and representing those who attacked Croatia (Cipek, 2009: 161). During the 1990s, almost 3,000 memorials in Croatia were damaged, removed or destroyed, most intensively during the war, while the state tolerated it (Banjeglav, 2012: 99–100). In Croatia, as Banjeglav shows, the destruction of memorials was an outcome of the intertwinement between the aggression against Croatia and erasure of the Partisan memory, identifying the armed forces of Milošević's rump Yugoslavia as a continuity with the Partisans (2012: 99–100).[3] In Serbia, where Milošević's regime did not want to give up the Partisan myth at any cost, memorials were not targeted.

The commemorations at the small-group and social level of anti-communist memory work revolve around several dates. First, alternative commemorations take place on the holidays that celebrate the People's Liberation War. The most prominent example is the annual church services for victims of communism in central Belgrade on the Day of Liberation on 20 October under the slogan "Bloody October". In the case of the Day of Uprising, two parallel commemorations take place in Bela Crkva, just a few hundred metres from each other. While one commemoration, far less festive than in the Yugoslav period, celebrates the beginning of the uprising, the other mourns the beginning of the tragedy of the civil war within the Serbian nation.

The Serbian Orthodox Church organises the commemoration of gendarmes Bogdan Lončar and Milenko Braković just across the street from the Partisan memorial complex in Bela Crkva. At the memorial unveiling in 2000, local priest Ljubomir Ranković repeatedly cried: "May it never happen again that a Serb kills a Serb", while hundreds of gathered people unanimously answered: "May it never happen again!" (Misojčić, 2000). On the 75th anniversary of the uprising in 2016, the Church added a memorial plaque, underlining: "Never again!". In addition to the Yugoslav holidays as occasions for parallel commemorations of Partisans' victims, religious holidays dedicated to remembering the dead, such as the Orthodox All Souls' Days (*Zadušnice*), also serve as the days of commemorations. The Lisičji Potok commemoration takes place on the All Souls' Day.

Commemorating the Chetniks

Chetnik commemorations mark the beginning and the end of the movement, taking place in May when Mihailović established the Chetniks on Ravna Gora and on the anniversaries of his arrest and execution in March and July respectively. State officials rarely attend these commemorations. If they do, it is an individual act and not an expression of the state support. Among the more dominant political parties, the consequent direct support for the Chetnik commemorations comes only from the SPO. Reflecting the hegemonic narratives, the commemorations of

Dragoljub Mihailović honour the Chetniks as both victims of communism and the first resistance guerilla in Europe. The national reconciliation narrative is a dominant theme that underlies these commemorations.

A church memorial service is a common form of commemorating the death of Dragoljub Mihailović in Serbia, Republika Srpska and emigration. The Serbian Orthodox Church does not initiate the requiems, but local Ravna Gora Movement chapters, political parties like the SPO or Dveri and other groups organise them. Since 1992, the Association of Members of the Yugoslav Army in the Homeland 1941–1945 (*Udruženje pripadnika JVuO*) has organised the largest memorial service for Dragoljub Mihailović in St. Sava Church in Belgrade. These religious services for the repose of Mihailović's soul take place around 17 July and their size varies.

While the memorial service in Sokobanja, for instance, gathers four people, St. Sava Church in Belgrade is full every year and a bishop and several other priests perform the requiem (*Nezavisne novine*, 2012; *Sokobanja – internet portal grada*, 2010). Patriarch Irinej officially blessed the service for the 70th anniversary of Mihailović's death in 2016 (*Danas*, 2017). Some services, such as by Dveri or the Ravna Gora Movement, honour the Romanov family in a parallel fashion, executed on the same day in 1918. Mihailović does not only represent the symbol of the suffering of the entire Chetnik movement but of the Serbian nation. As Bishop Arsenije explained in his 2017 requiem for Mihailović, "the shot in his brave soldier's chest was the shot at the truth, justice, bravery, persistence and loyalty to the fatherland and Orthodox faith" (*N1 Srbija*, 2016).

The annual Ravna Gora gathering is the central Chetnik commemoration. From an attempt to put up a memorial plaque in 1990, it grew to an attendance of thousands of people already in 1992 when the monument of Dragoljub Mihailović was erected. Organised by the SPO, it revolves around the interpretation of 13 May 1941 as the beginning of the uprising against the occupation, framing the Chetniks as the first anti-occupation guerilla in Europe. Another underlying theme is national reconciliation:

> From this place, where the immortal Serbian Čiča, General Draža Mihailović, took his famous Ravnogorci to the glory, we are sending a message of love to our brothers, all Serbs. We are sending the message of love to the whole world. The SPO, as a party of reconciliation, as a party of forgiveness, as a party of peace, will release doves from this place as a symbol of peace.
> (speech by Aleksandar Čotrić: *Srpska reč*, 1996: 17)

Ravna Gora Memorial Complex also encompasses a church and a memorial home. Intended as a library, an exhibition hall and a space for events, the memorial home was finished in 2000. The SPO asphalted the roads leading to the site. Only in 2005, the Ravna Gora gathering was officially supported by the Government of Serbia. However, the SPO was behind this support, with all three ministers who attended its members. No other officials were present. Nevertheless, because of their high positions, the state support for the Ravna Gora gathering faced harsh critiques in Serbia and especially in Croatia, resulting in Croatian president Stjepan Mesić cancelling his diplomatic visit to Serbia (Mautner, 2005).

The parallel commemoration by Ravna Gora Movement resembles the dual image of the Chetniks and divisions in this memory community. From the village Ba, the Ravna Gora Movement moved their commemoration to the central memorial complex, but they hold it a few days earlier. While the SPO put up the flags of western Allies of the Second World War, the Ravna Gora Movement celebrates Russia, commemorating the Romanovs and with guests from Russia and eastern Ukraine. The alternative gathering is not only smaller than the SPO one, but it is also more staunchly nationalist and militant. With uniforms, military hierarchy, line-ups and marching in columns, Ravna Gora Movement commemorations rather resemble a paramilitary exercise than a commemorative practice. Besides having a strict military codex, the Movement also involves an official chaplain and nurtures close relations with the Serbian Orthodox Church.

As a place of birth of the Chetnik movement, Ravna Gora is the central location that embodies their memory. However, it does not attract many visitors outside of the commemorations in May and July. When I visited Ravna Gora in September 2017, I was the only visitor. Once we left the main road near Mionica, the road that the SPO asphalted in the 1990s became bumpy and full of holes all the way to the memorial site. At the memorial home, Vladan Radosavljević, the son of Chetnik commander Miloš Radosavljević, who acts as a host at Ravna Gora, welcomed us with a table of tacky and overpriced Chetnik souvenirs, from fur hats to books by Vuk Drašković.

A wax figure of Dragoljub Mihailović decorates the exhibition hall still showing the 2012 exhibition "Ravna Gora Youth". In addition to that, the lists of all Chetnik units and their members cover walls, the project of the Chetnik veteran association carrying the title "65 Years Since the Victory Against Fascism". Another exhibition, portraying the history of the SPO in photographs, decorates the entrance hall with the souvenir table. While celebrating Dragoljub Mihailović and the Chetnik movement, the memorial complex at Ravna Gora emphasises the SPO as the main keeper of their memory. However, when the two commemoration days in May and July pass, Ravna Gora is abandoned and does not look like the epicentre of the hegemonic discourses about the Second World War in Serbian society.

Numerous Chetnik memorials exist in Serbia, many of which have been built in the post-Milošević period. Only a few memorials stand at central locations in towns, so, unlike the memorials on church property, they have the support of local municipalities. In Ivanjica, the birthplace of Dragoljub Mihailović, his statue has stood in the town centre since 2003, facing a Partisan memorial. A shop selling local and Chetnik souvenirs, is situated in Mihailović's family home nearby. The guides of the local tourist office bring visitors to the statue (*Politika*, 2009). A local Ravna Gora Movement revealed the monument, together with émigré Chetnik veterans who financed it. The local *Ravnogorci* had struggled for a memorial since the late 1980s, but the municipality rejected them (Šaponjić, 2003). After 2000, the political climate changed in favour of the Chetniks. The revealing of the monument marked the 100th birthday of Mihailović as General Draža's Days and had the support of mayor Milovan Marković of the DS and the Christian Democratic Party. Mihailović's grandson Vojislav represented the SPO.

In the town of Lapovo, the municipality erected the 4.6-metre tall statue of Mihailović in 2006, on the main square that was named after Mihailović a year before. Initiated by mayor Dragan Zlatković from the SPO, the Ministry of Culture and Diaspora supported the statue as well as donations from emigration from the United States (*B92*, 2006). Around 1,500 people gathered for the memorial uncovering, wearing Chetnik symbols and accompanied by loud Chetnik music and souvenir stands. The SPO officials gave speeches and two bishops performed the service for repose of Mihailović's soul. Unlike Ivanjica, this memorial was only possible because of the SPO dominance in the local municipality and ministries. Both cases illuminate the decisive role of the emigration in financing memorials.

The high number of Chetnik memorials across Serbia is not unique in the post-Yugoslav context. Namely, the Slovenian memory landscape has gone through a more profound transformation since 1991, with more than 200 memorials dedicated to Slovene collaborators erected by 2017 (Luthar, 2018: 34). As Oto Luthar explains, "there is at present almost no place without a memorial or a 'parish plaque' dedicated to the *Domobranci*" (2012: 887). Similarly to Serbia, memorials in Slovenia contain religious symbolism and take a central position at parish cemeteries. Honouring the collaborationist and anti-communist Slovene Home Guards, the memorials particularly commemorate those who had died in combat during the war or had been killed by the Partisans in its aftermath. They also promote the narratives of national trauma and reconciliation. As opposed to Slovenia, where Luthar counted 230 memorials built between 1991 and 2017, the Chetnik memorials in Serbia and Republika Srpska have proved to be very difficult to map and count.[4] This is not only because they are unofficial bottom-up initiatives, as in Slovenia, but also because these initiatives are often informal, do not use any platforms to promote their activities and rarely attract media attention.

The struggle for recognition

Anti-communist memory work from below relates to discontent with the attempts of revising the Second World War and dealing with the communist past. The failure of the state-sanctioned investigative commissions to, first, find the burial site and remains of Dragoljub Mihailović, and, second, to exhume and mark all postwar gravesites, is particularly frustrating. The importance of the gravesites to the actors of the small-group and social level is evident in their self-initiated involvement in both commissions. Furthermore, the *U ime naroda* exhibition and judicial rehabilitation gave these actors another opportunity to participate in the institutional level of memory work in the hope of recognition of their claims. The 1944 Commission, the exhibition and rehabilitation processes illuminate the informal nature of state-sanctioned memory politics and the mobility of actors up and down the hierarchy of memory work.

Perceiving the 2004 recognition of Chetnik veterans as antifascists and the possibility of rehabilitation as a step towards coming to terms with the totalitarian past, the actors of memory politics from below see them as insufficient and consider recognition in the public space in the form of a memorial as most relevant. Most of them agree that a memorial in central Belgrade, dedicated to all victims of communism, would be appropriate. The Association of Victims has been struggling for a memorial for years, however, without success. They have submitted several requests and petitions to the City of Belgrade but the city authorities have done nothing except inform them that a central location was out of the question. In 2017, they united with other actors in the Lisičji Potok memorial initiative.

For the SPO and the Chetnik veteran association, a memorial at Ada Ciganlija would make more sense because Mihailović had been executed there. Occasional commemorations at Ada took place already in the early 1990s. In 2011, the SPO organised a memorial service for Mihailović there, placing a wooden cross where Mihailović had allegedly been executed. Continuously advocating for a full memorial, they commemorate Mihailović there every year.

In addition to a memorial in Belgrade, the Association of Victims strives for marking all gravesites in Serbia in a modest way, with concrete crosses that would not be expensive or stolen. Because many locations of the postwar gravesites are used for diverse purposes today, they consider marking them to be the least the state and city authorities could do to pay respect to the dead. While they had advocated for the exhumation of all graves the 1944 Commission had found before, they do not expect all of them to be exhumed anymore:

> When it comes to rehabilitation, I think that's being done correctly. We do not have any objections there, we do not have any critiques about courts or their efficacy. It used to be very slow, but it is much better now, it is completely correct. Our objections are, however, first, that relation to the victims of the postwar repression in general. First, the attitude towards exhumations of victims, we consider it normal, we do not expect 215 mass graves to be exhumed. We know the state cannot do it, financially or anything. But, we think that at least each larger city in Serbia should exhume at least one bigger grave, each of them has seven or eight graves. So, exhuming at least one larger grave in every city in Serbia, that's the most elementary, minimal thing even for a state poorer than this one of ours. Not to stop exhuming 215 graves after the first one. This is a huge shame, a civilisational shame.
>
> (Đurić, 2016)

The 1944 Commission was the first state initiative that mobilised the actors from below. A few hundred people went through it, volunteering and queuing to have their stories finally heard. *Novosti* feuilleton about the postwar retribution had already empowered them by giving space to their testimonies, and the Commission came as a symbol of the long-awaited state support. The family members provided materials and testimonies that were crucial for mapping the gravesites. What they received in return was recognition through inclusion in the database and information and support in requesting rehabilitation of their family members. Their testimonies were not met with critique or suspicion.

These actors became even more actively involved once the 1944 Commission decided to present the results of their work in the form of an exhibition. The Commission had gathered enormous amounts of materials and contacts with the family members of the people executed in the postwar period or persecuted later. Even though the other Commission members were not in favour of this idea, Cvetković decided to make the results and materials the Commission had generated available to the public, as an exhibition and documentaries. Cvetković initiated *U ime naroda* as "a people's project" (Cvetković, 2016).

Joined by Zorica Marinković, he lobbied for support for the exhibition. They chose the Museum of History of Serbia as the former OZNA and Central Committee of the Communist Party headquarters, but they needed external funding. The Ministry of Culture financed the project only partially. Most funding came from private donors with a personal connection to the postwar repression, including Jelisaveta Karađorđević, emigration and Slobodan Homen's family law firm (Cvetković, 2014: 156). Descendants of the executed after the liberation or families whose property had been confiscated or nationalised featured prominently on the lists of sponsors and supporters.

That the exhibition "came from the people" was especially important for Cvetković, downplaying the institutional support. Although the exhibition had notable media support and both Serbian and foreign cultural institutions as partners, those involved prefer the image as coming from below, from the people. For them, the exhibition represents a way of fighting against the dominance of the Yugoslav memory culture. In Cvetković's words, the exhibition was "a struggle for freedom" (Cvetković, 2016).

Because of the regular events, the exhibition became a gathering point for the anti-communist memory community of Belgrade. The exhibition spoke to them and about them, involving them directly through stories about them and their family members. As such, *U ime naroda* became a commemorative act that lasted for months and went beyond the mere representations of the past depicted in the exhibition itself. Through the numerous lectures by Cvetković, the life of the exhibition extended and travelled throughout Serbia.

Even though the 1944 Commission and the exhibition represent the institutional level of memory work, they are equally bottom-up. This is evident in the establishment of the *U ime naroda* association that incorporated the Association of Victims, with Srđan Cvetković as president. They were also the main organiser of Lisičji Potok commemoration, a compromise the anti-communist non-state actors had to make as central locations in Belgrade remained unavailable, even for temporary memorials. Taking the matter of commemorations into their own hands quite literally, the lack of institutional support pushed them to organise themselves to clean the area together, making space for the commemoration and provisional memorial.

The symbolic power

The holders of symbolic power with influence and authority are crucial for anti-communist memory work, giving this sphere more legitimacy and meaning. These actors are prominent intellectuals, the Serbian Orthodox Church and the Karađorđević family, who usually endorse, attend and actively participate in commemorative practices like as Lisičji Potok. These practices usually involve a combination of a religious memorial service for the dead, a public reading of a letter by Aleksandar Karađorđević and a speech by one of the present intellectuals.

The intellectuals involved in anti-communist memory work are "the Belgrade critical intelligentsia" and others prominent as dissidents already during late state socialism (Dragović-Soso, 2002), together with a group of historians who gained prominence after 2000. They are politically heterogeneous but united in the anti-communist consensus. Many members of the Belgrade intelligentsia had been active in the historical debates of the 1980s. For instance, poet Matija Bećković had talked about the civil war between the Chetniks and Partisans and the need for national reconciliation already in 1985, a few years before it became the central theme of a part of the political opposition (Dragović-Soso, 2002: 101). On Victory Day over Fascism in 2014, Bećković revealed the plaque to victims of communism in his former high school in Valjevo (Vujanac, 2014b).

The liberal critics of Yugoslav state socialism such as jurist Kosta Čavoški or literary theorist Nikola Milošević took part in anti-communist commemorative activities from the early 1990s. After 2000, they supported every initiative to rehabilitate the defeated political and military personalities of the Second World War, individually and as the members of the Serbian Liberal Party (from 2010 under the title the Serbian Liberal Council). Their activities are not limited to appearances at commemorations and the mechanism of judicial rehabilitation is the result of their efforts.

The aforementioned historians from the Institute of Contemporary History, namely Srđan Cvetković, Bojan Dimitrijević, Kosta Nikolić and Momčilo Pavlović, are important actors within the anti-communist memory community. This is due to their role as mnemonic agents of state-sanctioned memory politics, such as investigative commissions, rather than their academic publications. Dimitrijević and Nikolić were involved in the Lisičji Potok initiative. Cvetković, who mobilised the Association of Victims for all initiatives he is involved in, who leads the *U ime naroda* group and has a vast number of devoted followers, is an especially relevant actor. His lectures across Serbia can also be considered acts of commemoration where the local anti-communist memory communities come together.

The royal support

Since the early 1990s, members of the Karađorđević family have supported the commemorations of defeated Second World War forces and victims of communism. Their support varies from attending commemorations to sending a letter to be publicly read to the audience, as Aleksandar Karađorđević usually does. His letters have been read at Ravna Gora gatherings and at the commemoration in Lisičji Potok. Jelisaveta Karađorđević endorsed and attended the Lisičji Potok commemoration and the *U ime naroda* exhibition that she partially financed.

As the political opposition was forming in the early 1990s, Aleksandar II Karađorđević (officially Crown Prince of Yugoslavia) and his family started increasingly appearing in media. They commented on the situation in Yugoslavia, condemned communism and discussed their possible return and reestablishment of the monarchy. In 1991, Aleksandar visited Serbia for the first time, greeted by thousands of people chanting "We want the king!" People such as Matija Bećković, Nikola Milošević and Dušan Kovačević were in the official delegation waiting at Belgrade airport. The regime did not prevent the visit and mass gatherings, ignoring it instead (Milošević, 1991).

During the 1990s, the Karađorđevićs kept relations to Milošević's political opposition. Already in 1991, Vojislav Koštunica spoke in favour of a parliamentary monarchy, promising to hire legal experts to find a solution to the issues of citizenship, property and status of the dethroned royal family (Milošević, 1991). When the 1947 Decree that stripped the royal family of their citizenship was abolished in 2001, they moved to Belgrade and still reside in the state-owned villas. They have never taken public functions. At the 2004 elections, Jelisaveta Karađorđević ran for president of Serbia, leading the "Initiative for a More Beautiful Serbia", but winning only 2.01 per cent.

In 2012, the Government of Serbia organised the exhumation, transport to Serbia and the reburial of the remains of the Karađorđevićs. The remains of Petar II Karađorđević, with his wife, mother and brother, were reburied in the family mausoleum in Oplenac, followed by the religious ceremony by Patriarch Irinej, attended by Tomislav Nikolić and Ivica Dačić, the prime minister at the time (*Blic*, 2013b). Around 10,000 people gathered. In their speeches, the patriarch, Nikolić and Dačić stressed that the burial in Oplenac is a symbol of Serbian unification and reconciliation. As a member of the SPS, Dačić called his presence a contribution to "the national reconciliation of those who were on the opposed sides in the past" (*Večernje novosti*, 2013). Tomislav Nikolić, who often appeared in the public with Aleksandar Karađorđević, argued that the Serbian nation could not allow themselves divisions and injustice:

> One of the conflicts emerged during the Second World War and had a tragic epilogue, whose consequences we can feel even today. The fights and persecutions continued even after the Second World War, and many citizens ended tragically in peace: some were executed without a trial, while for

some, the trial only served to deliver the sentence that was prepared in advance. That suffering was experienced by all our families, even the family of the freedom fighter and father of modern Serbia: Grand Vožd Karađorđe Petrović.

(*Večernje novosti*, 2013)

The Lisičji Potok memorial initiative is equally about the memory of victims of the postwar executions and of the interwar Kingdom of Yugoslavia. In 2009, a tip about the grave of Dragoljub Mihailović led *Novosti* journalist Milenko Kovačević, together with Srđan Cvetković, to Lisičji Potok (Kovačević, 2018). Weeds covered the area, but they discovered that the overgrown grass hid a memorial fountain built for assassinated King Aleksandar in 1936. Soon afterward, Cvetković mobilised his contacts for cleaning up the area and they started with small-scale commemorations. Before the annual commemorations at this site, a small group gets together to clean the area around "the king's fountain" and prepare it for the commemoration. This is an unofficial initiative, affiliated with Cvetković, *U ime naroda* and the Association of Victims, who use social media to call for helpers. The memorial park plan involves the parallel restoration of the fountain and building a memorial to victims of communism (Plan projekta: Inicijativa za obnovu spomen-česme kralju Aleksandru i podizanje spomenika žrtvama revolucionarnog terora 1944, n.d.).

The Karađorđevićs are particularly important as representatives of the time before communism. Negative reinterpretation of socialist Yugoslavia intertwines with the positive view on the interwar Kingdom of Yugoslavia. In this view, it was the golden era when the Serbian elites, factory owners, wealthy peasants and other respectable people of Serbia proper had thrived, while the communists had been rightfully outlawed. The communist takeover of power, hence, represents the destruction of the Serbian elites and the defeat of the social and political order of the Kingdom of Yugoslavia. The Karađorđevićs, as the Serbian dynasty that had ruled the Kingdom, were themselves defeated and lost all privileges and property, symbolising the downfall of once thriving society. As the successors of the Royal Yugoslav Army and endorsed by the king and the government, the Chetniks of Dragoljub Mihailović represent an inseparable aspect of this paradigm.

Religious dimension of anti-communist memory work

Religious memorial services for victims of communism and the Chetniks became a common commemorative practice of anti-communist memory communities in Serbia in the early 1990s. This tendency continued in the period after 2000. The support of the Serbian Orthodox Church involves several, usually parallel forms: the clergy attends commemorations, they perform religious services for the dead or give blessings to memorials and the Church provides space for memorials and plaques at its property, which is why many memorials to victims of communism are in churchyards or cemeteries.

Similarly to the Karađorđevićs, the commemorations are an opportunity for the Church to emphasise its victimhood under communism within the broader narrative of victimisation of the Serbian nation. In March 2016, when Srđan Cvetković gave a lecture in Vaznesenjska Church in central Belgrade, not only did the lecture specially underline the suffering of the Church and clergy after the 1944 liberation, but it began with an introduction by the priest and a joint prayer. This church, together with the Saint Sava Temple in Belgrade, is one of the locations of the annual memorial service for victims of communism that takes place as an alternative to the official celebrations of the Liberation Day. On the evening of Cvetković's lecture, they collected signatures for "the first memorial to victims of the communist regime in Belgrade" (Facebook, 2016). The lecture room of the parochial house was full. The predominantly older audience, dressed up for the occasion, was excited every time Cvetković referred to concrete cases, numbers and communists, growling insults at the expense of the perpetrators.

There are two memorial churches in Serbia directly linked to the commemorations of victims of communism and the forces who had lost the Second World War. The church at Ravna Gora was built in 1998, with its foundations blessed by the clergy in 1996 as the "foundations of reconciliation" (*Večernje novosti*, 1996: 10). Bishop Lavrentije blessed the church when it opened on 9 May 1998, financed by the SPO in Serbia and abroad (*Srpska reč*, 1998: 19). Memorial services for the dead have been a regular and central part of the Ravna Gora gathering.

In Šabac, a memorial church to victims of communist terror stands next to the bridge where post-liberation executions had taken place. The bridge is the symbol of the victimhood of the city under communism. Local citizens, including judge Gojko Lazarev, initiated and financed the church. Immediately after 2000, the day of the city and Liberation Day, 23 October, was abolished and the references to the People's Liberation War and socialist Yugoslavia disappeared from the street and institutions names. Church service for victims of the liberators replaced liberation celebrations. The local group behind the initiative for the memorial church is The Bridge (*Most*), gathering descendants of victims and other anti-communist-oriented locals. In addition to commemorations, they organise events dedicated to the theme of communist crimes.

Memorialisation of the postwar victims in Serbia is unthinkable without the Serbian Orthodox Church. Most memorials and plaques stand at church property, at cemeteries or in church gardens. The clergy blesses those outside of the church property, such as the memorial cross in Lisičji Potok or Mihailović statue at Ravna Gora. Locals usually initiate the church memorials and plaques that commemorate and mourn the local Chetnik men and others killed at the end of the war. A cross is a very common form of memorial for victims of communism and religious symbols and inscriptions are very characteristic. In Mionica near Valjevo, locals placed a marble cross at a non-investigated gravesite of 2000. The inscription dedicates the memorial to the Serb martyrs who had lost their lives just because they had thought differently but includes the Chetnik slogan "For King and Fatherland" and a prayer. The Valjevo bishop uncovered the plaque (Vujanac, 2014a).

The SPO and local Ravna Gora Movement erected a large concrete cross at Bagdala above Kruševac already in 1989, with the inscription "Lord, Have Mercy on us". In 2001, they added a memorial plaque dedicated to the Chetnik movement below the cross. The plaque depicts the portraits of Dragoljub Mihailović and Dragutin Keserović with the inscription, dedicating it specifically to the Chetnik corps:

> Memorial to the fighters of the Rasinski and Jastrebački Corps who fell in the fight against the occupiers of the Kingdom of Yugoslavia, as well as to all reputable Serbs and priests who were killed at this spot by the communist villains after the war together with the Chetniks. Grateful Serbhood.
> (*Večernje novosti*, 2009)

The highest clergy supports memorials to the most notorious Chetnik units and commanders, even when they are not in relation to the victimhood under communism. For instance, the village of Štitkovo near Nova Varoš sparked media attention in Serbia and Bosnia because of a Chetnik memorial complex erected by the descendants and other locals. In 2010, they erected the statue of Vuk Kalajitović, the commander of the Mileševo Chetnik Corps, whose units had killed thousands of Muslims of Sandžak and east Bosnia in 1943. After the final defeat of the Chetniks, Kalajitović did not want to surrender and continued fighting the Partisans. Surrounded by the Yugoslav Army, he committed suicide in 1948. Mileševo Bishop Filaret blessed his statue. The bishop blessed the memorial plaques with the names of 108 Kalajitović's Chetniks added in 2013, on 28 August, celebrated as the Dormition of the Mother of God (*Blic*, 2013a).

Notes

1 Matija Bećković is a nationalist poet who was very prominent in the 1980s political debates. Dušan Kovačević is a renowned playwright, member of the Crown Council and a public proponent of rehabilitation of Milan Nedić. Jelisaveta Karađorđević is the daughter of Prince Pavle Karađorđević.
2 Additionally, Kosta Nikolić and Dušan Bataković are listed as board members of the initiative for memorial. See Plan projekta: Inicijativa za obnovu spomen-česme kralju Aleksandru i podizanje spomenika žrtvama revolucionarnog terora 1944. (n.d.). Available at: http://online.fliphtml5.com/dskws/nvlt/#p=1 (accessed 8 October 2018).
3 In the post-Tuđman period, many memorials have been reconstructed or renovated.
4 The Institute of Culture and Memory Studies of the Academy of Sciences and Arts in Ljubljana made an online collection of memorials dedicated to the Home Guards, including descriptions and photographs, as well as the typology of them as outlined by Luthar: www.arzenal.si/sobe/zbirke/domobranski-spomeniki.

Bibliography

B92 (2006) "Spomenik Draži Mihailoviću". Available at: www.b92.net/info/vesti/index.php?yyyy=2006&mm=07&dd=11&nav_category=12&nav_id=204290 (accessed 19 October 2018).

Banjeglav, T. (2012) "Sjećanje na rat ili rat sjećanja? Promjene u politikama sjećanja u Hrvatskoj od 1990. godine do danas". In Karačić, D., Banjeglav, T. and Govedarica, N (eds), *Revizija prošlosti: Službene politike sjećanja u Bosni i Hercegovini, Hrvatskoj i Srbiji od 1990. godine*. Sarajevo: ACIPS, pp. 91–163.

Bieber, F. (2006) *Post-War Bosnia: Ethnicity, Inequality and Public Sector Governance*. Basingstoke, New York: Palgrave Macmillan.

Blic (2013a) "Otkriveno spomen obeležje Kalajitovim saborcima". Available at: www.blic.rs/vesti/drustvo/otkriveno-spomen-obelezje-kalajitovim-saborcima/yqw5m52 (accessed 13 October 2018).

Blic (2013b) "Posmrtni ostaci četvoro Karađorđevića sahranjeni na Oplencu". Available at: www.blic.rs/vesti/tema-dana/posmrtni-ostaci-cetvoro-karadordevica-sahranjeni-na-oplencu/bek9gm0 (accessed 11 October 2018).

Cipek, T. (2009) "Sjećanje na 1945: Čuvanje i brisanje". In Bosto, S. and Cipek, T. (eds), *Kultura sećanja: 1945. Povijesni lomovi i svladavanje prošlosti*. Zagreb: Disput, pp. 155–167.

Cvetković, S. (2014) *U ime naroda! Politička represija u Srbiji 1944–1953*. Belgrade: Evro-Đunti, Udruženje političkih zatvorenika i žrtava komunističkog režima.

Cvetković, S. (2016) Interview by author.

Danas (2017) "Parastos Draži Mihailoviću u ponedeljak". Available at: www.danas.rs/drustvo/parastos-drazi-mihailovicu-u-ponedeljak/ (accessed 14 October 2018).

Dragović-Soso, J. (2002) *Saviours of the Nation?: Serbia's Intellectual Opposition and the Revival of Nationalism*. London: C. Hurst & Co. Publishers.

Đurić, S. (2016) Interview by author.

Facebook (2016) "Predavanje dr Srdjana Cvetkovića: U Ime naroda! represija u Srbiji 1944–1953". Available at: www.facebook.com/events/610174512483468/ (accessed 12 October 2018).

Hram svetog oca Nikolaja – Draževina (n.d.) "Spomenik Draži Mihailoviću". Available at: www.drazevina.rs.sr/spomenik.htm (accessed 5 October 2018).

Inicijativni odbor (2017) Plan projekta: Inicijativa za obnovu spomen-česme kralju Aleksandru i podizanje spomenika žrtvama revolucionarnog terora 1944. Available at: http://online.fliphtml5.com/dskws/nvlt/#p=1 (accessed 8 October 2018).

Kovačević, M. (2018) "Kako smo tragajući za Dražinim grobom, pronašli zaboravljenu česmu kralja Aleksandra". In *Urbane strane*. Available at: http://urbanestrane.rs/2018/10/13/kako-smo-tragajuci-za-drazinim-grobom-pronasli-monumentalnu-cesmu-kralja-aleksandra/ (accessed 17 October 2018).

Lazarev, G. (2016) Interview by author.

Lovrić, I. (2013) "Ivan Ivanović: Prolazio sam kao u Stradiji". Available at: www.novosti.rs/vesti/kultura.71.html:415663-Ivan-Ivanovic-Prolazio-sam-kao-u-Stradiji (accessed 9 October 2018).

Luthar, O. (2012) "FORGETTING DOES (NOT) HURT. Historical Revisionism in Post-Socialist Slovenia", *Nationalities Papers* 41(6): 882–892. DOI: org/10.1080/00905992.2012.743510.

Luthar, O. (2018) "Post-Communist Memory Culture and the Historiography of the Second World War and the Post-War Execution of Slovenian Collaborationists", *Politička misao : časopis za politologiju* 55(2): 33–49.

Mark, J. (2010) *The Unfinished Revolution: Making Sense of the Communist Past in Central-Eastern Europe*. New Haven: Yale University Press.

Matijević, V. (2003) "Uklonjen spomenik Draži Mihailoviću u Brčkom". *Nezavisne novine*, 5 December.

Mautner, T. (2005) "Zagreb und Belgrad streiten über Tschetnik-Bewegung". Available at: www.dw.com/de/zagreb-und-belgrad-streiten-%C3%BCber-tschetnik-bewegung/a-1589409 (accessed 14 October 2018).
Milošević, M. (1991) "Povratak Karađorđevića – Aleksandrova formula". Available at: www.vreme.com/cms/view.php?id=1115805 (accessed 11 October 2018).
Misojčić, D. V. (2000) "Dve Srbije na sto metara". Available at: http://arhiva.glas-javnosti.rs/arhiva/2000/07/10/srpski/R00070905.shtm (accessed 14 October 2018).
Mondo Portal (2017) "Zemlja se neće smiriti do sahrane žrtava komunizma". Available at: http://mondo.rs/a1054319/Info/Drustvo/Beograd-spomenik-zrtvama-komunistickog-terora-Lisicji-potok.html (accessed 8 October 2018).
N1 Srbija (2016) "Služen parastos Draži Mihailoviću". Available at: http://rs.n1info.com/a177754/Vesti/Vesti/Sluzen-parastos-Drazi-Mihailovicu.html (accessed 14 October 2018).
Nezavisne novine (2012) "Biljana Plavšić na parastosu Draži Mihailoviću: Uvijek sam mu se divila". Available at: www.nezavisne.com/novosti/bih/Biljana-Plavsic-na-parastosu-Drazi-Mihailovicu-Uvijek-sam-mu-se-divila/150179 (accessed 14 October 2018).
Nježić, T. (2014) "Dragoslav Mihailović o izložbi 'U ime naroda': Hoće da se i sada osećamo krivima". Available at: www.blic.rs/kultura/vesti/dragoslav-mihailovic-o-izlozbi-u-ime-naroda-hoce-da-se-i-sada-osecamo-krivima/j116zzr (accessed 10 October 2018).
Politika (2009) "Ivanjičani čekaju čiča Dražu". Available at: www.politika.rs/scc/clanak/91194/Ivanjicani-cekaju-cica-Drazu (accessed 19 October 2018).
Samardžić, M. (2016) Interview by author.
Šaponjić, Ž. (2003) "Dobro došao, đenerale". Available at: http://arhiva.glas-javnosti.rs/arhiva/2003/04/29/srpski/R03042803.shtml (accessed 19 October 2018).
Sokobanja – internet portal grada (2010) "Parastos đeneralu Dragoljubu Mihailoviću Draži". Available at: http://sokobanja.eu.org/index.php?option=com_content&task=view&id=1182&Itemid=1 (accessed 14 October 2018).
Srpska reč (1996) "Temelj crkve, temelj pomirenja". 20 May.
Srpska reč (1998) "Spisak priložnika crkve Velikog velikomučenika Georgija na Ravnoj gori". 14 May.
Telegraf (2013) "Drašković: Srpski partizani razapinju Dražu i 'Ravnu goru'". Available at: www.telegraf.rs/vesti/880135-draskovic-srpski-partizani-i-dalje-razapinju-drazu-i-ravnu-goru (accessed 30 September 2018).
Večernje novosti (1996) "Pomen đeneralu". 12 May. Belgrade.
Večernje novosti (2009) "Pod Bagdalom Vaskrs pomirenja". Available at: www.novosti.rs/vesti/naslovna/aktuelno.290.html:243241-Pod-Bagdalom-Vaskrs-pomirenja (accessed 13 October 2018).
Večernje novosti (2013) "Sahrana Karađorđevića: Dinastiji mir, a Srbiji pomirenje". Available at: www.novosti.rs/vesti/naslovna/reportaze/aktuelno.293.html:435791-Sahrana-Karadjordjevica-Dinastiji-mir-a-Srbiji-pomirenje (accessed 11 October 2018).
Vujanac, P. (2014a) "Mionica: Spomenik žrtvama komunističke vlasti". Available at: www.blic.rs/vesti/srbija/mionica-spomenik-zrtvama-komunisticke-vlasti/8j93b82 (accessed 13 October 2018).
Vujanac, P. (2014b) "Otkrivena spomen-ploča žrtvama komunista u Valjevskoj gimnaziji". Available at: www.blic.rs/vesti/srbija/otkrivena-spomen-ploca-zrtvama-komunista-u-valjevskoj-gimnaziji/wrz7w05 (accessed 13 October 2018).
Živanović, K. and Jeremić, V. (2014) "Antić: Pravedno da obeležja četnika budu na paradi". Available at: www.danas.rs/drustvo/antic-pravedno-da-obelezja-cetnika-budu-na-paradi/ (accessed 17 March 2018).

8 History, memory and law[1]

In 1990, the SPO and the Liberal Party from Valjevo planned to stage a public trial against Josip Broz Tito and renew the processes against Dragoljub Mihailović and Milan Nedić. Tito would face trial as guilty for "the democratic, economic, moral, and overall collapse of the whole country" (Srpski pokret obnove, 1990: 27). With Vojislav Šešelj as the prosecutor, "writers, poets, publicists, memoirists, composers, sculptors, movie artists and journalists, who made him an overman and a god" would represent the defendant, or their work would speak for them if they refuse to testify or are no longer alive (Srpski pokret obnove, 1990: 27). The initiators wanted to repeat the 1946 trial against Dragoljub Mihailović, considering the process manipulated because it declared "the first guerilla fighter of occupied Europe a fascist quisling, an extraordinarily honourable and honest man a war criminal and a great Serbian and Yugoslav patriot a traitor" (Srpski pokret obnove, 1990: 27). To prove its manipulated nature, they wanted Miloš Minić, the prosecutor in Mihailović's trial, to present the identical case at the repeated process. The initiative got as far as publishing the indictment and mainstream newspapers portrayed it either negatively or satirically (Ninković-Džafo, 1990: 6; Šešelj, 1990).

Going further, *Pogledi* magazine filed numerous lawsuits against former Partisans, including the prominent persons such as Milovan Đilas. Courts eventually rejected all these cases. Regardless of the efforts proving unsuccessful, editor Miloslav Samardžić continued prosecuting the Partisans in Serbia throughout the 1990s. The idea emerged because many people had contacted the magazine, looking for their family members and friends. The letters section of Pogledi, printed on four pages, was a space for expression of grief about communism, often involving testimonies about the postwar executions and praises of the Chetnik movement. The lack of reaction to their texts especially frustrated *Pogledi* and Samardžić:

> Another paradox appeared before us. The shocking facts presented to the public, not in low circulation, did not spark any reaction. The graves remained not commemorated and unmarked, perpetrators unpunished, even though there is no statute of limitations for a war crime. This is the obvious

explanation of the paradox: all these big crimes were committed by the communists who rule this country even today.

(Samardžić, 1991)

There were attempts at the rehabilitation of victims of communism in the early 1990s as well. The Chamber of Lawyers established a commission to compose a proposal for rehabilitation of Serbian lawyers persecuted for their political convictions from 1941, to reconcile them regardless of their side during the civil war (Vujović, 1992a: 24). Opposed to only communist celebrations of their profession and commemorations dedicated only to the lawyers who had fallen as Partisans, the leading jurists of Serbia sought to morally rehabilitate their colleagues, as the Serbian legislation did not allow them to take the matter to a court. While gathering facts about their colleagues' destinies, the Chamber's goal was inscribing the names of the lawyers on the same plaque that commemorated the lawyers who had fought in the Partisans (Vujović, 1992b). In the post-Milošević period, the Chamber of Lawyers would play an important role in the context of judicial rehabilitation.

Although the discourses of victimhood under communism had surfaced in the public in the 1980s, the debates on the postwar retribution and political repression in state socialism intensified in the 1990s. Several initiatives went beyond the public revelation of crimes of communism and took legal action to indict or rehabilitate perpetrators and victims respectively, aiming at righting the wrongs of the distant past. These attempts by Milošević's opponents could not bring results because the legislator and judiciary did not allow such initiatives at the time. This changed drastically after the overthrow of Milošević.

Law has been an especially important instrument of state-sanctioned memory politics in Serbia since 2000. While no Yugoslav communists faced trials, the law enabled the rehabilitation of those considered victims, including the defeated military forces like the Chetniks. The first of the legislative steps regulating historical reinterpretation were 2004 changes to the Veteran Law that granted the same rights to Chetnik and Partisan veterans (*Službeni glasnik Republike Srbije*, 2004). Furthermore, the parliament passed two Rehabilitation Laws in 2006 and 2011, introducing the possibility of judicial rehabilitation for persons executed, sentenced or deprived of any rights for political and ideological reasons (*Službeni glasnik Republike Srbije*, 2006; *Službeni glasnik Republike Srbije*, 2011).

The mechanism of judicial rehabilitation usually seeks to accommodate victims of unfair trials by revising court processes or declaratively and collectively rehabilitating victims of political persecution and repression. In the case of Serbia, however, the legislation enables rehabilitation of the Axis collaborators and forces who bear responsibility for mass atrocities during the Second World War. The only criterium that makes it possible is that political and ideological grounds also played a role in their later judicial or extrajudicial punishment. The processes that have taken place according to the Rehabilitation Laws since 2006 have questioned and annulled postwar sentences or declarations of persons in

History, memory and law 131

question war criminals, collaborators, traitors and people's enemies, rewriting the history of the Second World War and socialist Yugoslavia in the courtroom.

While other post-Yugoslav countries with opposed sides during the Second World War, such as Croatia and Slovenia, also offered pensions and veteran rights to the Partisans' enemies, the mechanism of judicial rehabilitation of victims of communism, as defined and implemented in Serbia, represents a specificity in the post-Yugoslav space. In the same way as in Serbia, the national reconciliation narrative often fostered attempts to equalise the Partisans and collaborators across the post-Yugoslav space. In Croatia, the act enabling pensions for the veterans of the armed forces of the Independent State of Croatia and the Croatian Home Guard was passed in 1993, giving them the same status as the Partisans, suiting the policy of national reconciliation advocated by Franjo Tuđman.

In addition to veteran pensions, a few people have been rehabilitated in Croatia, including Archbishop Alojzije Stepinac, whose 1946 sentence was annulled in 2016, a week after his nephew had submitted the request for revision (Županijski sud u Zagrebu, 2016). Rehabilitation of Stepinac and others was according to the Criminal Procedure Act that regulates the possibility of rehabilitation in many countries, including Serbia. However, no post-Yugoslav country has adopted legislation specifically directed at the rehabilitation of victims of communism, while encompassing the entire Second World War. In fact, the Serbian legislation does not specifically refer to communism at all, but its understanding and implementation is exclusively for this purpose.

Equalising the Chetniks and the Partisans

The Veteran Law changes equalised the Chetniks with the Partisans by granting the Chetniks the same rights that the People's Liberation War veterans had enjoyed for decades. The law regulates the conditions for the fulfilment of rights for "veterans of the People's Liberation War, the participants of the earlier wars for liberation of the country, their family members, disabled soldiers and users of family disability allowance" (*Službeni glasnik Republike Srbije*, 2004: 2). According to the law, the veterans of the People's Liberation War are "the men who joined the People's Liberation War before 1 January 1944 and the women who joined it before 1 January 1945" (*Službeni glasnik Republike Srbije*, 2004: 2).[2]

The second paragraph introduces "the status of the veteran of the People's Liberation War to the members of the Yugoslav Army in the Homeland and Ravna Gora Movement, starting from 17 April 1941 and ending with 15 May 1945" (*Službeni glasnik Republike Srbije*, 2004: 2). The law grants the same status and rights for the Chetniks as for the Partisans by merging them together in the category of the People's Liberation War veterans. Furthermore, the law introduces *Ravnogorska spomenica 1941* medal for the people who joined the Chetnik movement between 17 April and 31 December 1941, making it equal to the already existing *Partizanska spomenica*, that the Partisans who joined the movement in the first year of the war received.[3] The value of *Spomenica* is not only symbolic, but it also involves an additional financial grant, paid monthly (*Službeni glasnik Republike Srbije*, 2004: 24).

In a similar manner to the museum exhibition changes in Serbia, the original legal text that regulated Partisans' rights remained intact, now with the Chetniks added to it, making the narrative of two antifascist movements of Serbia official. The law uses the term People's Liberation War (*Narodnooslobodilački rat, NOR*). This is not a neutral reference to the Second World War in Yugoslavia but relates specifically to the Partisans and their struggle for the liberation of the country from the occupiers and their domestic helpers. This was the term used in socialist Yugoslavia strictly for the Partisans, crucial for its legitimacy. The 2004 law, thus, merely added the Chetniks to the existing legislation without changing any details, even if contradictive. In this way, the Yugoslav Army in the Homeland became fighters of the People's Liberation War which is the name they had never used to describe themselves and the name of their enemies they had fought against for most of the wartime.

Furthermore, the law encompasses all fighters of the Yugoslav Army in the Homeland and Ravna Gora Movement from the capitulation of the Kingdom of Yugoslavia on 17 April 1941 to the official end of the Second World War in Yugoslavia on 15 May 1945. There are no limitations to this frame, even though the Chetniks of Dragoljub Mihailović were no longer on the Allied side after 1943. The law also does not consider other specificities of the Second World War, such as the 1944 amnesty given to the Chetniks and, more importantly, collaboration and war crimes. The legal text introduced pensions and other benefits for any Chetnik veterans or members of their families without any limitations.

The last revision of the Veteran Law before 2004 had been in 1990. The common practice of pasting the Chetniks onto the existing frameworks made the Chetniks and Partisans two antifascist movements of Serbia, without revising the leftovers of the veteran legislation of socialist Yugoslavia that are still present in the Serbian law. Some of the outdated characteristics of the Veteran Law concern the veterans who are either certainly not alive or whose families have already claimed their benefits. Such is the section that regulates the extraordinary monthly allowance that concerns the veterans of the People's Liberation War and earlier wars as well as families of fallen soldiers. However, the article further lists "the participants of the Spanish Civil War, the pre-war revolutionary movement, members of armed formations in antifascist struggles in other countries and persons who distinguished themselves in active work abroad on helping the People's Liberation War" (*Službeni glasnik Republike Srbije*, 2004: 23). Adding the Chetnik veterans, the main purpose of the law revision, the authors of the amendments overlooked the outdated articles that really need revision.

Implementation of the Veteran Law

The Veteran Law outlines the procedure for the fulfilment of all veteran rights and benefits from the local to state level. The recognition process for the status of a veteran of liberation wars starts at the municipal level, where the municipal authority for matters of veterans and invalid protection decides about each individual case. If the request involves the extraordinary monthly allowance, the municipal authority forwards the request to a commission formed by the government assembles to make the final decision. This commission consists of seven members, four delegated by the parliament and three by SUBNOR, and they do not have to provide justification of their positive or negative decision (*Službeni glasnik Republike Srbije*, 2004: 43).

In June 2005, the government issued a special degree dealing with the process of fulfilment of rights and benefits of the Chetnik veterans and their families (Vlada Republike Srbije, 2005, from now on: the Decree). The Decree goes into detail defining veteran rights and status, the war invalid status, rights of family members and *Ravnogorska spomenica*. The document in parallel involves the revisionist interpretation of the Second World War that negates the war between the Chetniks and Partisans and serious limitations that made it impossible for the Chetniks to gain recognition and benefits.

This document gives rights to the Chetniks who had been punished for their wartime activities, even if sentenced to death. The regulation of the war invalid status includes people who were wounded, injured or got ill while participating in the People's Liberation War as members of the Yugoslav Army in the Homeland in the period between 17 April 1941 and 15 May 1945 (Vlada Republike Srbije, 2005: 3). This paragraph also refers to the Chetniks who fought against the Partisans, as it encompasses the entire war period. Another paragraph of the governmental Decree directly addresses the participants in the People's Liberation War who died following the consequences of a wound, injury or a disease during a court trial indicting them for belonging to the Chetnik movement, and those sentenced to death because of that belonging (Vlada Republike Srbije, 2005: 4).

Both the Veteran Law and the Decree, that further regulates its implementation, leave out the war between the Partisans and the Chetniks, which would make it more difficult to place them shoulder to shoulder as veterans of the People's Liberation War. The formulation of these documents does not problematise collaboration and crimes against civilians, nor does it exclude those responsible and sentenced for both from fulfilling veteran rights. Therefore, even the persons sentenced to death for their wartime activities can still be considered for the status of veteran of the People's Liberation War.

The Decree, however, involves limitations and requirements that make it almost impossible for the Chetniks and their families to acquire the status of People's Liberation War veterans. Namely, they must prove their belonging to the Chetnik movement in the defined period with written evidence and "qualified witnesses". The qualified witnesses are "People's Liberation War veterans whose status has been determined and who were active in the same unit or in the same territory of operations" (Vlada Republike Srbije, 2005: 11). In other words, two Partisan veterans must submit letters of support. This is a consequence of the already mentioned SPS amendments to the Veteran Law.

With submission of requests possible only until 31 July 2007, no Chetnik veteran managed to receive recognition. In addition to the mandatory letters of support, the involvement of three SUBNOR representatives in the main decision-making body was the key factor why the Chetnik veterans could not acquire the status of the People's Liberation War veterans. SUBNOR boycotted the entire process and made the formation of the commission impossible. For Boro Ercegovac, the president of the Belgrade branch of SUBNOR, there was nothing wrong with this attitude of SUBNOR, but he argued that the state had also lost interest for the Chetniks:

> Is it possible that the state now suddenly accepts that SUBNOR ignores and makes the work of that commission impossible? I don't think so. If Toma Nikolić, the Chetnik duke who has not given up his title, is leading this country, why is he not pushing for it? He is in the position to push it. However, it seems that even the state has given up all that.
>
> (Ercegovac, 2016)

The Chetnik veteran association, created in 2004 as the Chetnik counterpart to SUBNOR with direct links to the SPO, could not do anything to change the veteran rights procedure. They share the opinion of SUBNOR that the state has given up on them and their cause. Formal and symbolic recognition of the Chetniks as a resistance movement, but without any practical consequences, is very significant grounds for frustrations, especially because there are always fewer veterans alive. They blame both the unwillingness of the state authorities and SUBNOR:

> The basic problem is that, if someone wants to prove that he was a member of the Ravna Gora Movement, he must have documents, some sort of military identification and evidence to prove it to the municipal commissions for the recognition of that status or have someone who already has that status who would testify for him. Taking into consideration that only members of the Partisan movement have acquired that veteran status, not a single *Ravnogorac* has managed it, it is mission impossible. They also do not possess any documents because everything was either destroyed by the communist authorities or by the members of those formations after the Second World War so that it could not be used against them as grounds for an indictment. The law remained a dead letter and the commission that was supposed to make decisions on the second level has never been formed. If a municipal commission does not solve the request by a Ravnogorac positively, to recognise his status, he has right to go to the republic commission, but the law regulates that SUBNOR sends its members to that commission and SUBNOR refuses to do so, so it is all one vicious cycle which we have been circulating in for 12 years.
>
> (Čotrić, 2016)

Rehabilitation legislation

Two Rehabilitation Laws, passed in 2006 and 2011, defined the possibility of judicial rehabilitation for those considered victims of politically motivated repression or violence. Political and military actors who were on the defeated side of the Second World War comprise only a small part of the processed case but the rehabilitation practice according to both laws has encompassed prominent historical actors too. It is no longer possible to file new requests as the 2011 Rehabilitation Law enabled the submission of rehabilitation requests for only five years, which ended in December 2016. Any still ongoing cases are based on the already submitted requests.

The 2006 Rehabilitation Law introduced potential rehabilitation for persons who "were deprived of life, freedom or other rights with or without an administrative or judicial decision on political or ideological grounds from 6 April 1941 to the date of the law's commencement" (*Službeni glasnik Republike Srbije*, 2006). The legal text is a two-page document with nine paragraphs that define rehabilitation in the broadest terms without instructions or limitations regarding implementation. The general formulation of the legal text does not provide any definition of "political and ideological grounds" or limitations concerning persons eligible for rehabilitation.

The time frame the law refers to begins with the outbreak of the Second World War in Yugoslavia and encompasses the entire wartime period. This is a specificity of Serbian law in the postsocialist context, because similar legislation in other countries, such as the 1991 Rehabilitation Laws in Ukraine and Russia, takes the establishment of a communist regime as the starting point. The Serbian laws refer to the period before Yugoslav state socialism, which is also wartime. As the law includes individuals who lost their lives without an administrative or judicial decision but does not omit persons who lost their lives in combat, there is space for rehabilitation of those who died in combat against ideologically and politically opposed enemies.

Explaining the time frame and speaking in the name of the law, proponents in the parliament, Zoran Stojković from the DS, emphasised that they took the date of the beginning of the Second World War because the system fell apart on that day and totalitarian elements appeared (*Stenografske beleške sa druge sedmice prvog redovnog zasedanja 04.04.2006*, 2006). Parliament members criticised the law heavily. Nevertheless, the law was adopted with the majority of votes, with 107 voting in favour and 20 against out of 129 present parliament members (*Privremene stenografske beleške sa druge sedmice prvog redovnog zasedanja, 9. dan rada*, 2006).

One month after the adoption of the law, the Supreme Court discussed its implementation and decided that non-adversarial proceedings should be applied in rehabilitation processes (Vrhovni sud Srbije, 2006). Non-adversarial proceeding allows only one side in the process so only plaintiffs who submitted a rehabilitation request and a judge or a court council that presides over and makes the final decision are involved. As the Rehabilitation Law allows complaints only in the case of a rejected rehabilitation request, and no right to complain if the court judgement is positive, the Supreme Court took it as a sign of no intent to involve the potential opponents in the process. The involvement of additional people in rehabilitation processes, besides plaintiffs, was left to courts to allow for the purposes of the determination of truth (Vrhovni sud Srbije, 2006).

Plaintiffs should submit a description of persecution or violence including information that should help identify the persons and events more closely, in those cases where no documentation and evidence exist or are available (*Službeni glasnik Republike Srbije*, 2006: 3). Although the court could request documentation from archives and other institutions, they usually base their decisions on the claims and documentation submitted by plaintiffs. As the law introduced the possibility of rehabilitation in very general terms without any instructions for the judiciary, courts had to resort to learning by doing, which resulted in highly inconsistent and diverse implementation in different regions in Serbia.

In conversations with several judges, I learned that judges from different courts exchanged materials and insights immediately after the 2006 law came into force, to create a common court practice in the field of rehabilitation. Nevertheless, this rather informal cooperation only included judges with a special interest in rehabilitating victims of communism and did not result in a country-wide consistent court practice or guidelines for the judiciary. The outcome of a rehabilitation process depends on the nature of the case but also on the court responsible for it. In some cases, courts require only one document from archives, very often a death certificate, and base their decision solely on the plaintiffs' claims. In some instances, however, the court requested a series of documents from multiple institutions, usually in the case of more prominent persons (Baković, 2011).

Another Rehabilitation Law was adopted in 2011, expected to improve the procedure and provide more guidelines for the judiciary. The law goes into detail regarding the conditions, procedure and consequences of rehabilitation, widening the reasons for persecution on "political, religious, national and ideological" grounds and if an administrative or judicial decision was against legal standards and human rights and freedoms (*Službeni glasnik Republike Srbije*, 2011). Unlike the 2006 law, which did not prescribe financial compensation and restitution, this law regulates conditions for restitution and compensation for those rehabilitated or their descendants.

The 2011 law includes important limitations about people who belonged to occupation and quisling forces and committed or took part in war crimes, who do not have the right to rehabilitation. This includes persons who died in combat as members of occupation or quisling forces. Furthermore, persons sentenced as war criminals by Yugoslav courts and other authorities could not be rehabilitated (*Službeni glasnik Republike Srbije*, 2011: 2). Although it is still possible to rehabilitate someone if the rehabilitation process determines that they did not commit war crimes, many Second World War actors who have been rehabilitated according to the 2006 law probably would not even have the right to rehabilitation under the 2011 legislation.

The 2011 law abandoned the use of non-adversarial proceedings and introduced the prosecutor's office as the other side in the courtroom. My observation of rehabilitation practice shows that the prosecutor is mostly passive during the process and their activity is based on filing complaints against the court decision, not proposing new documentation or inviting witnesses who would oppose the claims of plaintiffs during the initial process. Complaints repeated numerous times complicate and prolong rehabilitation processes and make it impossible for researchers and other interested parties to access court materials, as the judicial process must finish before public access is granted.

Most rehabilitation requests had been filed before the 2011 legislation. They are subject to the 2006 Rehabilitation Law, creating an overlap in court practice with cases under both laws heard at the same time. Additionally, the 2011 law enabled submission of rehabilitation requests only for a period of five years. This period ended in December 2016, leaving only the existing cases to be concluded. Regardless of occasional critical reactions that question the legitimacy of the completed prominent rehabilitation processes and taking into consideration the ten-year period from 2006 to 2016 to submit rehabilitation requests, it is highly unlikely that any new legislation regarding rehabilitation will appear in the future.[4] The still ongoing cases only represent an epilogue that brings this issue in Serbia to the end.

Telling histories in the courtroom

The issues that arise in the implementation of the Rehabilitation Laws come from their formulation, especially vague and problematic in the 2006 law. Several issues characterise the courtroom interpretation of history in rehabilitation processes: the prescribed conditions for rehabilitation, a very selective approach to documents and historical events, the participation of historians as expert witnesses and the issue of guilt and responsibility. Both the formulation of the legal text and dominant historical narratives are important factors that inform the rehabilitation practice and enable revisionist historical interpretation in courts. By transcending into a parallel political rehabilitation, judicial rehabilitation is a tool and a mirror of memory politics.

A specificity of both Rehabilitation Laws in the postsocialist context is that the legal text does not mention communism at all and does not define the law's purpose as specifically dedicated to victims of the communist authorities. Nevertheless, the implementation of the law and understanding of its purpose is only for rehabilitating victims of the Yugoslav regime before the end of the Second World War and in the immediate postwar period.[5] In practice, the persons rehabilitated are mainly those who were

> sentenced to prison, executed or punished by other means judicially or administratively by the organs of the People's Liberation Movement or communist authorities during and after the Second World War, declared as war criminals and traitors by the State Commission for the Determination of

the Crimes of Occupiers and Their Collaborators, or persons subject to repression without any formal decisions.

(Milošević, 2013: 34)

Although certainly in many cases it makes sense to question an accusation of collaboration or treason and there are numerous cases of too harsh punishment, the lack of limitations in the legal text enables rehabilitation of persons who indeed collaborated with the occupation or bear command or direct responsibility for mass crimes. Because of this, the mechanism of rehabilitation in Serbia does not merely represent rehabilitation of innocent victims of politically motivated repression, but it serves as the relativisation and justification of collaboration and crimes of the Second World War through law and courts. The judicial process exonerates historical actors of their responsibility and rehabilitates them, regardless of what they did during the Second World War, revising the war history.

Rehabilitation is achievable for victims of persecution or deprivation of rights based on political and ideological reasons, and religious and national grounds that the 2011 law added. A case can hinge on a judicial or administrative decision, but the lack of a documented decision does not represent an obstacle. The definition of rehabilitation is not only about persons sentenced in postwar trials or those who by public decrees are declared as enemies of the people. In the case of Dragiša Cvetković, the prime minister of the Kingdom of Yugoslavia who signed the Tripartite Pact, rehabilitated in 2009, the court in Niš took the 1945 Decree declaring him an enemy of the people as the grounds for the positive decision. The fact that the document central to the process did not exist at all, not having been preserved, did not hinder his case. When Pavle Karađorđević was rehabilitated in 2011 along with the entire royal family, it did not matter that there had been no arrest, trial or court judgement against any of them (Milošević, 2013).

The decrees of confiscation of property or declaration of someone as a people's enemy very often serve as the main and only reason for successful rehabilitation as victims of political and ideological persecution. In the case of Chetnik commander Nikola Kalabić, rehabilitated in Valjevo in early 2017, the judge rehabilitated him based on the posthumous court Decree of 1946 that ordered confiscation of property, evaluating the Decree as politically and ideologically motivated. The court proceedings discussed that Kalabić incurs command responsibility for terror against civilians in central Serbia, but the court judged it irrelevant to his rehabilitation. Similarly, there is no court verdict in the case of Milan Nedić, the collaborationist prime minister in occupied Serbia, but the ground for requesting rehabilitation was the 1949 Decree declaring him an enemy of the people and subsequent confiscation of property from his family.[6]

The historical interpretation that carries weight in rehabilitation processes emanate from a highly selective approach to, and interpretation of, history and historical records. This problem arises partly from the nature of non-adversarial proceedings in rehabilitation cases conducted under the 2006 law. However, the cases under the 2011 legislation are no different in this regard. The processes are not top-down, and the plaintiffs play the main role, but the legislation itself does not enable narratives opposing the dominant interpretation entering the courtroom. Following the vague legal text, plaintiffs submit their claims and supporting documentation but are also in charge of choosing and inviting witnesses and experts who would testify in favour of rehabilitation. This automatically leads to the interpretation of history that omits or denies all facts not speaking in favour of someone's rehabilitation.

In the most prominent cases, such as rehabilitation of Dragoljub Mihailović, Milan Nedić or Nikola Kalabić, plaintiffs invite historians as expert witnesses. They are professional and clearly positioned historians with interest in the case, providing more credibility to plaintiffs' claims even though their testimony does not necessarily correspond to known facts and historical records.[7] During the process of Mihailović's rehabilitation, the historians who testified as expert witnesses had been known proponents of his rehabilitation and positive reinterpretation of the Chetnik movement (except Branko Latas, who was included as a result of pressure by civil society). In the cases of Nedić and Kalabić, not that many professional historians openly and publicly support their rehabilitation but both cases included historians from state-funded institutions as expert witnesses, including the already mentioned anti-communist historians from the Institute of Contemporary History.

The legislation clearly does not take the Second World War and the question of guilt and responsibility as the subject of rehabilitation; instead, it concerns whether a person was "deprived of life, freedom or other rights with or without administrative or judicial decision" and on political or ideological grounds (*Službeni glasnik Republike Srbije*, 2006). However, courtroom debates in the most prominent cases discuss the Second World War to the extent that they marginalise the persecution of the person in question. The most prominent example of this problem is Mihailović's rehabilitation process, where expert witnesses discussed the wartime context, the question of collaboration and crimes and anti-occupation activities of the Chetnik movement. Although Mihailović's 1946 trial and court verdict constituted the foundations for his rehabilitation, the proceedings rather focused on the war and the nature of the Chetniks and Mihailović, portraying him as either unaware of or not responsible for collaboration and crimes and as the leader of the resistance movement.

In a similar manner, the Axis occupation of Serbia and the role of Milan Nedić and his authorities was the main theme of the process of Milan Nedić's rehabilitation. Some proceedings were hours-long discussions of the topics such as the economy in Nedić's Serbia. The plaintiffs and their witnesses promoted the portrait of Nedić as a saviour of Serbian refugees from the Independent State of Croatia, unaware of the Holocaust in Serbia, who justly persecuted communists for their revolutionary plans, because of which he eventually became a victim of the communist regime.

The emphasis on the wartime context gives more significance to judicial rehabilitation and makes it more than a correction of past injustice. Making a positive decision after such testimonies and proceedings, as is the case with

Mihailović, courts affirm revision of the Second World War and its actors. As the court rehabilitated Mihailović after a process based on discussion of the war, his rehabilitation is not only a revision of the 1946 trial and verdict, but also an official confirmation of a positive narrative about the Chetnik movement and judicial approval of the decades-long process of their revaluation in Serbian society. Much the same applies in other cases of military and political actors of the Second World War whose rehabilitation processes are a judicial aspect of the dominant politics of memory based on a revision of the Second World War and state socialism.

Although confiscation of property is often the main motivation for a rehabilitation request, plaintiffs usually have to submit a separate request to the Office for Restitution, upon a rehabilitation request or a successful rehabilitation process (*Službeni glasnik RS*, 2015). The Office can withdraw the rights to return of property and compensation in specific cases, even when a rehabilitation process is still ongoing. This is the case with both Nedić and Kalabić, while no one has ever submitted a restitution request for Mihailović. The absence of restitution or any financial compensation for descendants of military and political actors makes judicial rehabilitation essentially a symbolic and political mechanism. Moreover, while rehabilitation is often framed as a family matter seeking to return honour to a family name, individuals and groups with political interest in the outcome of a rehabilitation process join the descendants as plaintiffs. This further contributes to the characteristic of the practice of judicial rehabilitation as politically motivated.

Judicial abolishment of the uprising

In 2008, the Regional Court in Šabac rehabilitated gendarmes Bogdan Lončar and Milenko Braković as victims of persecution and violence, without any guilt and victimised primarily for the communists' goal of evoking fear among those who could oppose the Partisans and the Communist Party ideology (Okružni sud u Šapcu, 2008: 6). The court interpreted the killing of Lončar and Braković by Španac in 1941 as the beginning of the communist revolution under the cover of the struggle for liberation from the fascist occupation, the main goal of the Communist Party of Yugoslavia being the revolutionary takeover of power (2008: 6). The materials submitted by the plaintiff and expert opinion of Kosta Nikolić were the main foundations of the case.

The court council, presided by Gojko Lazarev, argued that the event in Bela Crkva was the beginning of the civil war in Serbia. According to them, the gendarmes were no occupiers or people's enemies, but representatives of the legitimate Serbian state authorities under occupation. Therefore, the rehabilitation of the gendarmes is the beginning of national reconciliation of the divided Serbian nation (Okružni sud u Šapcu, 2008).

The main and only plaintiff was Stana Munjić, a local journalist without any connection to the gendarmes or the case. Her main motivation for requesting their rehabilitation was the revival of the Day of the Uprising and the state support for it, as opposed to the holiday's official abolishment. Having seen two announcements about 7 July as the Day of the Uprising on websites of the Ministry of Defence and Ministry of Labour and Social Affairs, she was frustrated by their denial of the parliamentary abolishment of the holiday as the official acknowledgement that the event in Bela Crkva was "one of the forgeries of the official Serbian history after the Second World War" (Munjić, 2008). In her request, she further explains that she considers it "her obligation to contribute to preventing further deviation, because the manipulation of history continues and the state does not respect its own decision about quashing 7 July as the official holiday" (Munjić, 2008). She interviewed Lončar's widow, wrote a book and made two local television documentaries and submitted these materials to the court as the main evidence.

The case of Lončar and Braković was the first rehabilitation case with a historian invited to provide an expert opinion, that subsequently became a standard practice. In Šabac, it was judge Lazarev who asked Kosta Nikolić to write an expert statement about what happened in Bela Crkva in 1941. The court decision incorporated large parts of Nikolić's essay as arguments in favour of gendarmes' rehabilitation. For Nikolić, and the court council, the Partisans fought primarily for the revolution and seizure of power, rather than the liberation, which is why they predominantly attacked Serbian authorities at the beginning of the war:

> In the four years after the military defeat and occupation, the large overturn happened in the Serbian nation, society and the state. One whole world was destroyed with the replacement of Serbian royalists with Yugoslav communists, and the irony is that Germans had a small or almost no role in it.
> (Okružni sud u Šapcu, 2008)

In his expert opinion, Nikolić argues that the celebrations of 7 July in socialist Yugoslavia had the purpose of justifying the violent takeover of power by the Communist Party. He explains that "celebrating the murder of Lončar and Braković has left long-term negative consequences and obstructed the process of national reconciliation and overcoming the ideological division that are still strong" (Nikolić, 2008: 25). Rather than only providing an expert opinion on a historical event, Nikolić recommends rehabilitation of the gendarmes, arguing it will represent a great contribution to facing the totalitarian legacy in Serbian society, the legacy that is still a serious obstacle to the full modernisation and democratisation of Serbia (2008: 25).

The expert opinion of Nikolić is a great example of anti-communist revisionist historiography in Serbia. In addition to the negative interpretation of the Partisans, Nikolić perpetuates the hegemonic narrative of the civil war between the Chetniks and Partisans, blaming the communists for starting the conflict. The Partisans' central objective of revolution hindered the possibility of a united front with "the other antifascist group", the Chetniks. Interpreting the Chetniks as antifascist, Nikolić underlines their wide recognition and legitimacy as opposed to the illegitimate nature and goals of the communist-led Partisans, who Nikolić openly condemns (Nikolić, 2008: 17).

Having declared them victims of persecution and violence by communists, the case of Lončar and Braković is a rehabilitation process that aims at reinterpretation of history rather than concrete compensation for victims and their families. No restitution or compensation followed, as the plaintiff did not have any relation to the gendarmes, and no one requested recompense. With the discourses of victimhood, civil war and national reconciliation, the court case reflects the hegemonic discourses about the Second World War in Serbia and the actors promoting them. It also illuminates the inconsistencies of official memory politics. Although 7 July had ceased to be a public holiday in 2001, the state institutions regenerated it as the Day of the Uprising in the Second World War in 2013. Even though its status is not as important as it was before 2001, state and municipal representatives officially commemorate it in Bela Crkva. At the same time, gendarmes Lončar and Braković, rehabilitated as the first officially recognised victims of the civil war and communism, remain legally rehabilitated as such, but commemorated only through the anti-communist memory work from below.

Rehabilitation from below

Plaintiffs in rehabilitation processes in Serbia are usually people like Stana Munjić. They are not state actors and they can be family members, but they often do not have any direct relation to the people whose rehabilitation they fight for. The 2006 Rehabilitation Law states that "any interested natural or legal person" can file a request (*Službeni glasnik Republike Srbije*, 2006: 2). The interested person is not just any person, but broadly defined as someone with connections to the subject of rehabilitation and interest in the positive outcome of the rehabilitation process (Apelacioni sud u Beogradu, 2011). In the practice, it is usually descendants who act as plaintiffs, with the support of organisations and groups of the anti-communist memory community. As judge Lazarev summarises:

> An injured party files the request if it is about the violation of property, freedom or other rights, or their descendants, particularly in the cases of people extrajudicially executed or sentenced to death. Those eligible to request rehabilitation are either injured parties or their descendants, as well as some social organisations who deal with these issues. In our cases, it is most often the descendants because a long time has passed since human rights violations, the time has done its job and most of these people are no longer alive. The descendants file the documentation they have, and it is used as evidence. They can also propose evidence, such as witnesses or historical books. The court then obtains facts from historical archives, security services, other courts, state institutions, from all organs and organisations where the court assumes or the plaintiffs suggest the information could be found.
>
> (Lazarev, 2016)

The 1944 Commission played a significant role in aiding both the descendants who wanted to request rehabilitation of their family members, and Serbian courts, by providing information about the persons whose names are in the Commission database. When looking for documents about subjects of rehabilitation, judges not only investigate the online database, but they also officially contact the Commission, namely Cvetković, who was also active as an expert witness. The 1944 Commission advised people who approached them about their rights to rehabilitation and restitution. As the judicial system requires a person to be officially declared dead before potential rehabilitation, the plaintiffs submitted the signed statement saying something along the lines of: "We received the information about the victimisation from the State Commission for Secret Graves" (Ilić, 2016). In this way, judges knew they could contact the Commission for additional facts and documents and the Commission cooperated. This accelerated the rehabilitation processes.

> They asked. They asked what rights they have. My job was to ask them: Have you submitted a rehabilitation request? You have the right to request rehabilitation. My job was to inform them. Someone thinks this is not necessary, that their father or grandfather did not do anything bad, but some people wanted it to take off that sign they had been carrying for 50–60 years, that they were children of the stigmatised, of the stigmatised people's enemies. Sometimes it was about the victims themselves and sometimes about descendants who want to make some peace, inner peace with themselves.
>
> (Ilić, 2016)

Rehabilitation processes in Serbia always start as an initiative from below. More importantly, as the legislation defines the use of non-adversarial proceedings, the plaintiffs submit documentation and invite witnesses. While the court could obtain additional information or witnesses, there is usually no opposing side in the courtroom challenging the plaintiffs' claims. The state authorities offered the legal framework that enables the actors from below to use the courtroom as the platform for their interpretation of the past. Even though judges make the final decision, the courtroom represents the space where the voices of the plaintiffs must be heard and can present their claims to the wider public through the media presence at more prominent proceedings. Rehabilitation processes illuminate the motivation behind anti-communist memory work from below, which is rarely limited to righting the past wrongs inflicted upon family members and has an equally strong political dimension rooted in anti-communism.

The Association of Political Prisoners and Victims of the Communist Regime joined the plaintiffs in the cases of Dragoljub Mihailović and Milan Nedić. They perceived their involvement in the Mihailović's case as completely natural because of the savage, too harsh and revengeful character of the postwar punishment against the Chetniks that they saw as injustice. When it comes to Nedić, it is both his status among the interwar elite and the positive role during the occupation of

Serbia that matter. According to Slobodan Đurić, Milan Nedić wanted to help the Serbian people survive the war, he managed to pacify Serbia and enable normal life, as much as it was possible to do so (Đurić, 2016).

In addition to the Association of Victims, the rehabilitation processes bring together a large part of the anti-communist memory community of the small-group and social level of memory work. The Karađorđevićs were themselves rehabilitated and do not directly take part in other rehabilitation processes. However, the members of the Crown Council take an active role, such as Kosta Čavoški, who was one of the plaintiffs in Mihailović's case, and Zoran Živanović, the lawyer in the rehabilitation processes of both Mihailović and Nedić. Besides family members, the plaintiffs in the prominent cases such as these are the Association of Victims, Chetnik organisations, the Serbian Liberal Council and individuals such as Kosta Čavoški, Smilja Avramov and Miloslav Samardžić. As opposed to commemorations, anti-communist historians do not initiate rehabilitation processes, but support them through testifying as expert witnesses with statements exclusively favourable to the positive outcome of rehabilitation processes.

The representatives of the anti-communist memory community from below are generally satisfied with the rehabilitation practice, even though they criticise the difficulties of the financial consequences. The mechanism of judicial rehabilitation has a high symbolic value for plaintiffs and the anti-communist memory community. For family members, it is reinstating honour to the family name, usually intertwined with political rehabilitation of the person and the ideas they stood for. By requesting and shaping the rehabilitation process, the actors from below get the chance to move up the memory work hierarchy and promote their interpretation of the past, sometimes taking it to a large public stage. Because of the agency from below and their public dimension, the rehabilitation processes represent important contact zones of individual, small-group, social and institutional levels of memory work in Serbia.

Notes

1 Parts of this chapter were previously published as: Đureinović, J. (2018) "Law as an Instrument and as a Mirror of Official Memory Politics: The Mechanism for Rehabilitating Victims of Communism in Serbia", *Review of Central and East European Law* 43(2): 232–251. DOI: org/10.1163/15730352–04302005. Reproduced by permission of Brill.
2 The participants of the earlier wars for liberation of the country are those who participated in: the Balkan Wars, the First World War as members of the armed forces of Serbia and Montenegro, the 1917 Toplica Uprising, the volunteers of the Salonika Front of the Serbian army and the block of the countries Serbia and Montenegro belonged to.
3 The Decree describes *Ravnogorska spomenica* as

> round and with 49 millimetres diameter. The emblem of the Army of the Kingdom of Yugoslavia is placed on its left side, in silver colour, and on the right side, there is a figure of a soldier in gold, standing and holding the flag of the Kingdom of Yugoslavia in his left hand, while calling people to the fight with the right hand.

4 Besides civil society initiatives and critique that comes from historians, some state institutions also voiced criticism. For instance, the Office for Restitution submitted a proposal to the Ministry of Justice in 2016, arguing for revision of the Rehabilitation Law which would disable rehabilitation of members of the occupation forces but with no response so far. Restitution office representatives have raised numerous concerns in public about court practice and it usually rejects the restitution requests by rehabilitated persons, for whom evidence of collaboration in the Second World War exists.
5 There have been several cases of rehabilitation of persons imprisoned or persecuted in the later Yugoslav period, but this article limits its scope to actors of the Second World War. A few exceptions have appeared before Serbian courts requesting rehabilitation of Partisans killed during the war, such as the case of Roksanda Lišančić, a Partisan woman tortured and murdered by the Chetniks in 1943, who was rehabilitated in Čačak in 2013.
6 According to the official interpretation, Nedić committed suicide in prison while awaiting trial. There is no evidence proving he was killed. Prior to his death, the Commission for the Investigation of Crimes of Occupiers and Their Domestic Collaborators interrogated him, and these transcripts are available and have been published numerous times.
7 The recent development have been the involvement of historians who do not belong to the revisionist paradigm in rehabilitation processes, such as Milan Ristović in Nedić's rehabilitation process in 2017, who the Jewish community of Belgrade proposed, and Milivoj Bešlin in the repeated case of Nikola Kalabić, still ongoing at the time of completion of this manuscript in 2019.

Bibliography

Apelacioni sud u Beogradu (2011) "Rešenje Reh. 5/11". Available at: www.bg.ap.sud.rs/lt/articles/sudska-praksa/pregled-sudske-prakse-apelacionog-suda-u-beogradu/odeljenje-radnih-sporova/reh-5-11.html (accessed 22 October 2018).
Baković, N. (2011) "Using Local Archives for a Historical Reevaluation of Socialism. Examples of Bankrupted Factories' Collections and Rehabilitation Processes in Čačak Region (Serbia)", *Revista Archivelor* 2: 42–54.
Čotrić, A. (2016) Interview by author.
Đurić, S. (2016) Interview by author.
Ercegovac, B. (2016) Interview by author.
Ilić, M. (2016) Interview by author.
Lazarev, G. (2016) Interview by author.
Milošević, S. (2013) *Istorija pred sudom. Interpretacija prošlosti i pravni aspekti u rehabilitaciji Pavla Karađorđevića*. Belgrade: Fabrika knjiga.
Munjić, S. (2008) Zahtev za rehabilitaciju Bogdana Lončara i Milenka Brakovića. Okružni sud u Šapcu. Šabac: Okružni sud u Šapcu.
Nikolić, K. (2008) Istoriografski esej. Šabac: Okružni sud u Šapcu.
Ninković-Džafo, V. (1990) "Kad vojvoda sudi maršalu". *Borba*, 14 May.
Okružni sud u Šapcu (2008) "Rešenje, Reh. br. 11/08". Okružni sud u Šapcu.
Privremene stenografske beleške sa druge sedmice prvog redovnog zasedanja, 9. dan rada (2006). Available at: www.otvoreniparlament.rs/transkript/6327?page=1 (accessed 7 October 2017).
Samardžić, M. (1991) "Pravda". *Pogledi*, 2 August.
Šešelj, V. (1990) "Optužnica protiv Josipa Broza". *Pogledi*, 15 May.
Službeni glasnik Republike Srbije (2004) Zakon o pravima boraca, vojnih invalida i njihovih porodica 137/2004.

Službeni glasnik Republike Srbije (2006) Zakon o rehabilitaciji 33/2006.
Službeni glasnik Republike Srbije (2011) Zakon o rehabilitaciji 92/2011.
Službeni glasnik RS (2015) Zakon o vraćanju imovine i obeštećenju 72/2011, 108/2013, 142/2014, 88/2015.
Srpski pokret obnove (1990) "Zašto i kako suditi Josipu Brozu, Draži Mihailoviću i Milanu Nediću". *Pogledi*, 15 April.
Stenografske beleške sa druge sedmice prvog redovnog zasedanja 04.04.2006 (2006). Available at: www.otvoreniparlament.rs/transkript/6320?tagId=57703 (accessed 7 October 2017).
Vlada Republike Srbije (2005) "Uredba o načinu ostvarivanja prava pripadnika Jugoslovenske vojske u otadžbini i Ravnogorskog pokreta u oblasti boračko-invalidske zaštite". *Službeni glasnik RS*.
Vrhovni sud Srbije (Supreme Court of Serbia) (2006) Načelni pravni stav 629/2006.
Vujović, M. (1992a) "Advokatura uzvraća udarac". *NIN*, 1 May.
Vujović, M. (1992b) "Igre dostojanstvom". *NIN*, 15 May.
Županijski sud u Zagrebu (2016) *Presuda: Poslovni broj Kv-I-173/16*. Zagreb: Županijski sud.

9 Rehabilitation of Dragoljub Mihailović

On 14 May 2015, nine years after the rehabilitation request submission, judge Aleksandar Trešnjev read the decision to rehabilitate Dragoljub Mihailović in front of a full courtroom of the Palace of Justice in Belgrade. A loud applause and chanting greeted his announcement. The rehabilitation process ended in favour of Mihailović and annulled his 1946 indictment and conviction. Not being able to fit into the already crowded courtroom, around 400 people gathered in front of the Palace of Justice. Having heard about the decision, they yelled, hugged and chanted "Victory!". One of the plaintiff's lawyers, Oliver Antić, was welcomed with hugs as he stepped out of the courthouse and Vojislav Šešelj held a short celebratory speech in front of the court.

Numerous uniformed members of the Ravna Gora Movement gathered in front of the courthouse, some of them standing lined up holding flags of their municipal chapters. Members of Obraz, a far-right organisation banned in Serbia, supported them, as well as other smaller groups of a similar political provenance. Accompanied by a man dressed in a cassock and either completely uniformed or wearing a šajkača hat, they were singing a repertoire of Chetnik songs. They also yelled offense at the other, significantly smaller, group standing behind the police cordon holding a big yellow sign with a red inscription: "Death to Fascism, Freedom to the People". They tirelessly sang the Partisan and other revolutionary songs. A couple of Partisan veterans stood with them, shocked and disappointed about the court decision. The area in front of the Palace of Justice was swarming with journalists trying to capture the most interesting scenes.

The opposed gatherings in front of the court took place continuously during the rehabilitation process. Both sides chanted offense at each other and had to be kept separated by the police. The side of the street protesting Mihailović's rehabilitation had Yugoslav and antifascist flags, and different signs protesting rehabilitation and saying: "Stop Falsification of History", "Against Rehabilitation of Fascism" or "Rehabilitation of Criminals is Complicity". On the other side of the street, there were Chetnik flags, framed portraits of Mihailović and slogans. Often, the supporters of Mihailović's rehabilitation brought portraits of the prominent people of the 1990s war in Bosnia, such as Ratko Mladić and Radovan Karadžić. Further,

Women in Black had a slogan "Mihailović – genocide – Mladić" inscribed on a large black sign, also demonstrating continuities between the Second World War Chetniks and the 1990s Serbian forces.

The rehabilitation process of Dragoljub Mihailović involved opposing sides and all relevant actors of multiple levels of memory work. Even though the Serbian Orthodox Church, the Serbian Renewal Movement (SPO) and the Karađorđević family were not directly involved in the court case, they all publicly endorsed Mihailović's rehabilitation. It represents the culmination of the long process of political rehabilitation of the Chetniks and the judicial confirmation of their positive recasting. As such, the court case mirrors the hegemonic discourses about the Chetnik movement and their centrality for the dominant memory politics. Finally, the quest for Mihailović's grave that took place simultaneously to the rehabilitation process further contributed to his constant presence in media.

The court process involved two phases. During the first phase, that lasted until the involvement of Oliver Antić in 2013, the central theme of the court hearings was the Second World War and the nature of the Chetnik movement. The proceedings discussed the 1946 trial too, but it was more important to emphasise the positive image of Mihailović and prove that he had nothing to do with collaboration or crimes. Once Oliver Antić appeared, at the time as the advisor to Serbia's president, the emphasis changed. He drew attention to the procedural irregularities of the 1946 trial, dismissing everything discussed earlier in the process as protraction. Reflecting the hegemonic discourses, the first part of the rehabilitation process pivoted on the positive image of Mihailović as a leader of the internationally renowned resistance movement, while the latter part brought his image as a victim of communism to the fore.

Agency behind the court case

The rehabilitation process started when Vojislav Mihailović submitted a request for rehabilitation of his grandfather Dragoljub in 2006. Not to be mistaken for a family matter, the court case brought together the anti-communist memory community and had numerous organisations and individuals as plaintiffs, all with a vested political interest in the outcome of the process. While these actors did not cherish the identical image of Mihailović and the Chetniks, resembling the already addressed divisions, they united over negation of the Chetnik collaboration and Mihailović's victimhood. As opposed to Vojislav Mihailović who was invisible in the process, these groups and their lawyers stirred the judicial process and their voices dominated the public sphere.

The rehabilitation of Dragoljub Mihailović was the result of merging two parallel rehabilitation requests into one court case. Namely, two years after Vojislav Mihailović's request, legal scholars and practitioners composed and filed another claim for Dragoljub Mihailović's rehabilitation. The Serbian Liberal Party (later Council) instigated the request with 80 pages written by lawyer Zoran Živanović (Živanović, 2016). Serbian lawyers had strived for the rehabilitation of their colleagues from the early 1990s, initiating the rehabilitation of Slobodan Jovanović in 2004, two years before the first Rehabilitation Law, with Živanović as the key actor.[1] The Chamber of Lawyers had investigated the possibilities of rehabilitation of prominent historical actors prosecuted by the Yugoslav authorities before any legislation had existed to enable it, filing a request as soon as the Rehabilitation Act was passed in 2006. The case of Slobodan Jovanović was successful and was the official entrance of Zoran Živanović into the sphere of judicial rehabilitation.

Before becoming interested in rehabilitating Mihailović, the Chamber of Lawyers had officially rehabilitated Slobodan Subotić, the former Chamber's president and a lawyer in Mihailović's trial, sentenced for spreading enemy propaganda after he had protested the 1974 Constitution (Pejčić, 2009). In fact, the Chamber had rehabilitated Subotić already in the early 1990s, decorating him with a medal. The Rehabilitation Law introduced the possibility to rehabilitate him judicially too.

Considering the case to be far more extensive and complex than the others, Živanović refused to lead the rehabilitation of Mihailović at first when the Serbian Liberal Party contacted him. He started working at the International Crime Tribunal for the former Yugoslavia in the Netherlands and could not be present for a court case in Serbia (Živanović, 2016). However, two years later, when some prominent colleagues, such as Veljko Guberina, asked him to represent the Serbian Liberal Party in Mihailović's case again, he accepted. The list of plaintiffs grew quickly and eventually involved: the Association of the Members of the Yugoslav Army in the Homeland (from now on: The JVuO Association), the Association of Political Prisoners and Victims of the Communist Regime (The Association of Victims), Miloslav Samardžić and *Pogledi*, Kosta Čavoški and Smilja Avramov. The Ravna Gora Movement joined later. In 2013, the JVuO Association hired Oliver Antić.

The High Court in Belgrade merged the two cases in 2008 and the proceedings started in October 2010. For unknown reasons, Vojislav Mihailović did not attend court hearings nor did he hire a legal representative. In the court protocols of the first two proceedings, his name does not appear on the list of plaintiffs. Starting with September 2011, the protocols of every proceedings list all present plaintiffs and their representatives, with a note stating that Vojislav Mihailović had been regularly informed but did not attend (Viši sud, 2012a: 1). Every hearing ended with an explicit statement to send invitations to Vojislav Mihailović (and Smilja Avramov, who also did not attend any hearings). The familiar dimension of the process existed only formally, as Vojislav Mihailović involved only on paper.

The SPO and the Chetnik veteran association did not take part in the process:

> What rehabilitation? The rehabilitation before the court? No! We have rehabilitated him in other ways. We built him a memorial at Ravna Gora, we built a memorial home at Ravna Gora, we built the church of reconciliation at Ravna Gora. Throughout all these years while Yugoslavia was being

killed, we went to Ravna Gora with dozens of thousands of people and condemned the crimes of the Yugoslav People's Army, the ethnic cleansing, we wanted the evil to stop, the hatred not to win, the bloodbath to stop, convinced that the man of that mountain, if he had been alive, would have spoken the same way I did, as all of us did who went up that mountain in those years. Finally, I wrote the novel *Noć đenerala*. We pushed the law of equalisation of the two movements, both undoubtedly antifascist, and, in this way, we completed our work, we completed our mission of fighting for truth.

(Drašković, 2016)

The SPO greeted the positive outcome of the rehabilitation, with Vuk Drašković stating that it was more than the rehabilitation of the commander of the first anti-Nazi guerilla in Europe, it was the rehabilitation of his army and the whole of Serbia. According to him, the court finally removed the anathema that Serbia had been on the side of Hitler's Third Reich just because of supporting the leader of the anti-occupation movement and a minister in the Yugoslav government in exile (Glavonjić and Mihajlović, 2015). Similarly, the Serbian Orthodox Church greeted the process outcome. Patriarch Irinej declared that he believed the epilogue of Mihailović's rehabilitation would be the reconciliation of the Serbian people, overcoming the divisions into Partisans and Chetniks (Glavonjić and Mihajlović, 2015). Finally, the Karađorđevićs were interested in the court case, with Aleksandar Karađorđević attending the last proceeding and congratulating the plaintiffs. Calling for prayers for "everyone who lost their lives on all sides", he understood the court decision as crucial for national reconciliation (*Blic*, 2015).

Even though he did not take an active role in the process, Vojislav Mihailović was formally a plaintiff and publicly represented as such. The other plaintiffs judged his position and wanted the public to know who was really active. The Serbian Liberal Council published a statement about Mihailović's inactivity in the case, stressing that "all documents, materials, evidence and testimonies had been collected without the engagement and support of Vojislav Mihailović as a plaintiff and the Serbian Renewal Movement as a political organisation that had shown interest in the outcome of the rehabilitation" (*Nova srpska politička misao*, 2011). They called Vojislav Mihailović and the SPO to cooperate with the lawyers already successfully fighting for Mihailović's rehabilitation, but the SPO remained uninvolved and Mihailović kept ignoring the court invitations.

The court wanted to know the plaintiffs' motifs and they replied by emphasising the importance of reinstating the historical truth and justice harmed by the Yugoslav regime. Only Dušan Đukić from the JVuO Association, whose father Svetomir had been a Chetnik commander, had a direct connection to Mihailović, having met him as a child while retreating together through Bosnia at the end of the war (Viši sud, 2014: 3–4). The Serbian Liberal Council saw Mihailović's rehabilitation as rehabilitating the state and justice, preventing such trials from repeating (Viši sud, 2014: 3). The Council and Kosta Čavoški highlighted historical truth and justice as well as patriotism and the legitimate character of the Yugoslav Army in the Homeland as opposed to the violent takeover of power by the communists (Viši sud, 2014: 4). The Association of Victims, represented by Slobodan Đurić, also underlined the importance of finally revealing the truth about the unfair trial (Viši sud, 2014: 3). They all drew attention to Mihailović as a victim of communism.

The numerous plaintiffs in the rehabilitation process of Dragoljub Mihailović came from across the spectrum of nationalist politics, enjoying very different levels of prominence and influence in Serbian society. What they all have in common is not being state actors. Taking into consideration that Antić was the advisor of Tomislav Nikolić at the time, the president with the title Chetnik duke (*vojvoda*) that he earned during the war in Bosnia and did not want to give up, and that he crucially contributed to the ending of the process, a certain level of indirect state support existed. Certainly, no state authorities seriously opposed or condemned the process. While the court case reflects hegemonic discourses, it is also the expression of claims of the anti-communist memory community from below.

The Second World War in the courtroom

Partisan veteran Risto Vuković requested to participate in the process at its very beginning. During the Second World War, Vuković and his brother were wounded in combat against joint Chetnik and German units and Vuković fought remaining Chetniks for the State Security until the late 1950s (Viši sud, 2010a: 2). He wanted to reach moral satisfaction by testifying and hoped for an active role as the interested party because of his personal connection to the Chetniks and events of the Second World War. The interested party can become a plaintiff for having a personal, familial or historical connection to the subject of rehabilitation, having a stake in the process and its outcome or being affected by it. Even though lawyer Živanović protested, the court allowed Vuković to speak as a witness (Viši sud, 2010b: 2). However, Vuković refused to appear as a witness and insisted on being recognised as the interested party, which the court ruled out and then excluded him from the process (Viši sud, 2010a: 4).

Emphasising that a non-adversarial proceeding does not allow opposing sides, Živanović did not want Vuković to participate in any form. He claimed that Vuković could not possess information relevant for the process and that he wanted to present evidence of Mihailović's guilt, irrelevant for the rehabilitation case that focuses on the nature of the 1946 trial (Viši sud, 2010a: 2–3). Vuković's lawyer Bajo Cmiljanić argued with Živanović, claiming that his client was interested to prove that Mihailović's trial was not political by proving his guilt. Judge Aleksandar Ivanović finally agreed with Živanović that Vuković's potential testimony would not help determining the truth that only refers to the trial, while Vuković wanted to speak about the Second World War (Beta, 2010).

While the court rejected Vuković's involvement to prevent the discussion about the Second World War in the case, they allowed Milton Friend, a United States Air Force veteran, to testify about the wartime period on a parallel basis.

Explicitly emphasising that wartime is not a topic of discussion in a rehabilitation case, Živanović invited Friend who did not have any connection to the 1946 trial, but would speak in favour of Mihailović and his rehabilitation. The plaintiffs argued that he could testify because the American airmen rescued by the Chetniks in 1944 were not allowed to testify in the trial (Viši sud, 2010b: 2).

The purpose of Friend's testimony was the emphasis on the positive image of Mihailović. Speaking about the emergency landing in Serbia and being rescued by the Chetniks, Friend depicted Mihailović as "warm and kind", wearing the uniform similar to the one of British soldiers and speaking French (Viši sud, 2010b: 4). When the American airmen heard about the trial, they wanted to fly to Yugoslavia to testify. However, the State Department responded with: "the communists refused the pilots' testimonies because the criminal acts Mihailović had committed during the war were so hard that no testimony in his favour could change his guilt" (Viši sud, 2010b: 5). He ended his testimony with the story he had heard from George Vujnovich, an officer responsible for airlifting the airmen from Serbia, that he had offered Mihailović to take him to the Allied zone, "into the freedom", but Mihailović refused, saying his place is in Serbia with his people (Viši sud, 2010b: 5).

Historians as expert witnesses

The appearance of professional historians as expert witnesses informed the rehabilitation process of Dragoljub Mihailović. While Slobodan G. Marković opened the process, testifying as the representative of the 1944 Commission and Mihailović Commission, Bojan Dimitrijević and Kosta Nikolić spoke as historians of the Chetnik movement recommended by Marković as experts. Moreover, the involvement of Veselin Đuretić and Branko Latas was a result of efforts of different groups that supported or opposed Mihailović's rehabilitation. Even though none of historians who testified had engaged with the 1946 trial professionally, or with any other topics related to law, courts and postwar trials, they commented on Mihailović's trial. The Second World War constituted the central theme of historians' testimonies, in a similar manner like in Milton Friend's statement. They focused on the relations of the Chetniks with the occupation forces, the Partisans and the Allies and discussed Mihailović's responsibility for collaboration and, less prominently, crimes.

The only person who talked about the 1946 trial in detail was Slobodan Marković, whose testimony stretched over four court proceedings. In addition to presenting a great number of sources that demonstrated the Communist Party's manipulation of Mihailović's trial and its political nature, Marković offered the foreign, mostly British, perspective on the trial and the Chetniks (Viši sud, 2011, 2012d: 03, 2012e). He devoted attention to the role of the Central Committee and Politburo of the Party and OZNA in preparing the trial and observing the defendant's lawyers as well as the influence the state authorities might have had on Mihailović's defence. The court received a copy of the 2010 testimony by a guard who claimed that Mihailović was given strong alcoholic drinks on a daily basis, starting from 0.25 litre and ending with 1.25, which "could be significant for his ability to defend himself properly" (Viši sud, 2011: 7).

The largest part of Marković's testimony analysed the British documents on the trial, such as the diplomats' letters and the attempt by British officers to testify in favour of Mihailović. Arguing that the British witnesses would be crucial for the characterisation of the Chetniks at the trial, Marković claimed that the impossibility to testify was not only a procedural issue, but it influenced the trial's merit (Viši sud, 2011: 5). He assumed that the officers' testimonies would be along the lines of the 1947 secret report by William Mackenzie, depicting Mihailović as a great leader whose tactical, collaborationist initiatives were approved by the British government (Mackenzie, 2002; Viši sud, 2011: 6).

The British documents that the Mihailović Commission collected should prove that Mihailović had never collaborated with the German occupation and that he had condemned his officers for doing so (Viši sud, 2012d: 2–4). The western Allies decorating Mihailović was another argument serving this purpose. Furthermore, Marković argued that other defendants of the Belgrade Process emigrated to the West and were never extradited to Yugoslavia, which shows that "the western democracies did not accept their sentences as legitimate", even when relations with Yugoslavia were close (Viši sud, 2012d: 6).

Empty papers signed by Mihailović, that he had when he was captured, surfaced as important evidence. Marković and Kosta Nikolić found and photographed two of these papers in the Military Museum and one was published in Vladimir Dedijer's diary. They claim that the Yugoslav authorities used the papers for the falsification of letters to Ante Pavelić and Alojzije Stepinac and the 1941 order to commanders in Montenegro that called for the cleansing of non-Serbs. While the Commission proved that empty signed papers existed and were in the possession of the Yugoslav authorities, the Commission did not have any concrete evidence that someone filled these papers with compromising content to be used against Mihailović. Nevertheless, Marković concluded that these papers brought into question "the authenticity of every individual document that prosecutor Minić presented at the trial against Mihailović, with his signature and the stamp of the Command of the Yugoslav Army in the Homeland" (Viši sud, 2012d: 8).

Historians of the Second World War on the stand

The court invited Bojan Dimitrijević, Kosta Nikolić and Veselin Đuretić for "witness examination" about "circumstances of the criminal proceeding against General Mihailović and circumstances of their knowledge about sources, evidence and other materials used in the above-named criminal proceeding against Dragoljub Mihailović" (Viši sud, 2012a: 10). All historians' court statements were openly and entirely in favour of Mihailović's rehabilitation. Although they briefly referred to the trial, the wartime period was the focus of the hearings.

None of the expert witnesses was an expert for the Belgrade Process or has studied any other postwar trials, legal history and other law-related matters. As

Dimitrijević admitted, he was a historian and, as such, he studied the trial and was not familiar with the context, but he formed his opinion based on historical facts (Viši sud, 2012b: 3). Historians' testimonies did not reveal anything new or relevant about Mihailović or his trial that would drastically influence the general knowledge and justify the annulment of the sentence. They argued that the trial was rigged and involved many irregularities, emphasising the example of the forged letters to question the entire case against Mihailović.

The argumentation of falsification of letters in the court was vague and "methodologically rather unconvincing" (Dulić, 2012: 641). It is not entirely clear how the historians had reached the conclusion about the forgery. Dismissing letters served the purpose of exonerating Mihailović of all guilt, while portraying the Yugoslav authorities as manipulative and having to engage in forgery to construct the case against him. The historians reduced the issue of command responsibility and involvement in war crimes or collaboration on signing a written document (Dulić, 2012). The premise is that if Mihailović had not signed the order, he is not guilty, even though orders for mass atrocities are often not in writing and, finally, the atrocities did happen.

In his statement, Dimitrijević focused on the nature of the Chetnik movement and the Second World War, representing Mihailović as the leader of a resistance movement and representative of the legal and legitimate Yugoslav government who kept his anti-occupation attitude during the entire war and was perceived as an enemy by the occupiers (Viši sud, 2012b: 5–6). The only reason why Mihailović stopped his anti-occupation activities and turned against the Partisans were the drastic reprisals of the German occupation (Viši sud, 2012b: 6). For Dimitrijević, meetings with the German representatives were not acts of collaboration, because they did not result in cooperation, while some commanders outside of Serbia were forced to engage in tactical cooperation because of the relations with the Partisans (Viši sud, 2012b: 7–8):

> Mihailović did receive reports from his commanders about tactical cooperation with Italians in 1943 and with Germans in 1944, but he did not approve of that cooperation personally. He considered and explained in other papers that it represented the possibility for his movement to destroy the rival Partisan movement in certain territories, to prevent their progress and eventually obtain arms and ammunition. So, he did know but he did not deny it in the trial, that most of his commanders who got into combat with the Partisan forces mainly had to collaborate with the Italian, German or the forces of the Serbian government, or the Slovenian.
>
> (Viši sud, 2012b: 9)

Invited by "twelve Serbian organisations that cherish the tradition of Ravna Gora Movement and the Yugoslav Army in the Homeland", Veselin Đuretić did not testify, but submitted his expert opinion in writing, with the title *How and Why Was the Serbian Resistance Movement of Yugoslav Orientation Destroyed?* (Viši sud, 2012c: 2). His short talk was a nationalist political speech rather than a courtroom testimony, criticising the anti-Serbian character of the Partisans and socialist Yugoslavia (Viši sud, 2012c: 2–3).

Keeping his testimony rather short, Kosta Nikolić depicted the indictment and trial as rigged and moulded between Belgrade and Moscow. Underlining the illegitimate nature of the Partisans, he emphasised that Mihailović was the only legal representative of the Yugoslav Royal Army and that the Wehrmacht recognised his Chetniks as legal troops (Viši sud, 2012c: 4). In addition to the legitimacy status, the collaboration of the Partisans with the German forces in 1943 served as another point of differentiation with the Chetniks. Together with Dimitrijević, he wrote a report for the court, presenting the overview of anti-occupation activities of the Chetnik movement.

The redundancy of the historians' testimonies was in the representation of the image of the Chetniks from the British and German perspective, naturally, only focusing on what served their positive interpretation. In their statements, the Chetniks are a legitimate resistance movement representing the government in exile and prioritising the wellbeing of the Serbian people above everything, with the Partisans as the main source of their troubles. This directly implies a negative portrayal of the Partisans, bringing up their collaboration with the German forces as another irrelevant theme for the question of whether Mihailović's trial was politically and ideologically motivated.

The focus on the war and its selective interpretation delegitimise Mihailović's indictment, trial and conviction by demonstrating that everything the Yugoslav authorities had against Mihailović in the courtroom was false. Historians blamed the Yugoslav court for deliberately excluding facts that did not fit their interpretation of the Chetniks as traitors, war criminals and collaborators, with the goal of sentencing Mihailović to death. The paradox is that, by exclusively highlighting the positive qualities of Mihailović and the Chetniks and excluding all facts that did not accord with their positive image, these historians did the same as what they blamed the Yugoslav court for doing 70 years earlier.

Interventions against the process

The rehabilitation process drew public attention in Serbia and the post-Yugoslav space and generated both positive and negative reactions. The opposition within Serbia involved historians, SUBNOR and several leftist and liberal NGOs, informal groups and initiatives. In addition to regular protests in front of the court, groups intervened in the court case by submitting materials, proposing witnesses and filing a lawsuit against Dimitrijević and Nikolić. The court accepted the proposal to include historian Branko Latas as an expert witness, while other interventions were ignored and the lawsuit against historians slowed the process down rather insignificantly.

Only a small group of historians whose work reflects the hegemonic discourses is professionally involved in the institutional level of memory work. They testified during Mihailović's rehabilitation. Historians critical of dominant memory politics are rarely expert witnesses, because plaintiffs only invite those who would speak positively about the subject of rehabilitation. The inclusion of historians critical of revisionism only happens as a result of interventions, such as Milan Ristović in Nedić's rehabilitation process, or if the case is conducted according to the 2011 Rehabilitation Law. Many historians spoke out publicly during Mihailović's rehabilitation process and condemned it in the wider context of condemnation of historical revisionism but the debate between the two sides does not transcend into academic publications. One of the important reasons for the lack of academic polemic is that only a handful of historians in Serbia work on the Second World War and the postwar period and they are predominantly anti-communist.

SUBNOR and the Association of Antifascists of Serbia (*Savez antifašista Srbije*) continuously struggled for inclusion of their war interpretation in the courtroom and attempted obstructing the process. They did so separately, the Association of Antifascists having support of other NGOs.[2] While the court and plaintiffs ignored most of the materials they submitted, they succeeded in including Branko Latas as a witness.

In a similar manner as his opponents, Latas explained his was a legal layman and could not comment on legal questions related to the trial and focused on the wartime instead (Viši sud, 2013b: 2). However, he started with the statement that "the Chetnik movement of Draža Mihailović was collaborationist, by no chance antifascist" (Viši sud, 2013b: 2). His statement was similar to those by other historians, just coming from the opposite perspective and portraying Mihailović in an exclusively negative light, listing examples of collaboration and providing documents and a book to support his claims (Latas, 1999).

The plaintiffs did not welcome Latas's testimony. Lawyer Živanović interrogated him harshly. By the end of 2013, Oliver Antić was involved and wanted to prevent further hearing by Latas or any consideration of materials he submitted, because

> a witness in a civil court case, as in any other, can only testify about the circumstances personally and directly known to them, and submitting documents and testimonies of the other persons in a hearsay manner does not represent testifying, so such witness should not be allowed to testify.
> (Viši sud, 2013a: 2)

Even though lawyers obstructed his testimony, Latas represented the only critical view in the entire process, only enabled by interference from outside.

In July 2013, the Association of Antifascists, Women in Black and Helsinki Committee, among others, filed a lawsuit against Bojan Dimitrijević and Kosta Nikolić for "presenting incorrect information and false statements, although they knew, as historians, that their statements do not correspond to the truth" and committing a criminal act of forgery (Krivična prijava, 2013). They accused the historians of deceiving the court and leading it to allow the rehabilitation of "war criminal Dragoljub Mihailović" (Krivična prijava, 2013). The lawsuit listed all statements deemed problematic and references where the corrections could be found, such as an explanation that the allegedly forged letter to Pavelić was not used in the 1946 trial. The Public Prosecutor's Office rejected the case and the subsequent appeal.

Reactions in the region

When the court rehabilitated Mihailović in 2015, president of the Bosnian Social Democratic Party Nermin Nikšić called it "a slap in the face of all victims of Chetnik persecution", arguing that the same ideology inspired those who had tried to ethnically cleanse Bosnia of its Bosniak and Croatian population in the last war (Milekić et al., 2015). The process of Mihailović's rehabilitation sparked negative reactions across the post-Yugoslav space, particularly in Bosnia and Herzegovina and Croatia. Many political actors condemned it. While some demanded withdrawing ambassadors from Serbia, others criticised it but understood it as an internal issue of Serbia (*Nezavisne novine*, 2012; *Večernje novosti*, 2012; Spasojević, 2012). In Bosnia, Denis Bećirović of the House of Representatives of the Parliamentary Assembly, sent an open letter to Slavica Đukić Dejanović (SPS), the president of the Serbian parliament, arguing that the rehabilitation of the Chetniks and their ideology was not only an internal matter of Serbia but it legalised the Chetnik project of Greater Serbia at the expense of Bosnia, Croatia and Montenegro (*Nezavisne novine*, 2012).

The critics of Mihailović's rehabilitation process from Bosnia and Herzegovina had the 1990s wars as their lens, seeing the reinterpretation of the Chetniks as continuity and giving legitimacy to the crimes by the Serbian military and paramilitary forces in Bosnia in the 1990s. This view sees the rehabilitation of Dragoljub Mihailović as directly related to the Chetnik revival of the 1990s. Moreover, the crimes the Chetniks had committed against civilians in the territory of Bosnia and Herzegovina during the Second World War represent an important reference point. The commemorative and other activities of the Ravna Gora Movement in Republika Srpska make the issue of Chetnik rehabilitation a Bosnian matter too. Interestingly, the Serbian political parties or the officials of Republika Srpska did not comment on the court case.

While the negative reactions in Bosnia and Herzegovina symbolised the intertwined memory of the 1990s wars and the Second World War, the critique coming from Croatia reflects the wider process of the national fragmentation of memory that went hand in hand with the dissolution of Yugoslavia. The debates of the 1980s and the war in the 1990s remain relevant for the contemporary dynamics between Serbia and Croatia. The hegemonic discourses in both countries are to a large extent similar to the Second World War as the main object of revision, directly linked to delegitimising Yugoslavia. The war is observed through the national lens, as well as the prism of the postwar period. The defeated forces of the Second World War have become more suitable historical references than the anti-nationalist and communist-led Partisans.

As opposed to the similarities, the dominant narratives in Serbia and Croatia are incompatible and competitive, crucial for understanding negative reactions to rehabilitation of Mihailović.[3] The condemnation of the postwar executions and trials and rehabilitation of victims, regardless of their wartime activities, is a shared tendency. However, the nationalisation of memory implies the parallel negative interpretation of the trials in their own country as politically motivated and appalled condemnation of rehabilitation attempts in the neighbouring country. Even though the same Yugoslav courts operated in both countries, rehabilitation processes in the other countries are deemed historical revisionism. Hence, the same courts could be objective when sentencing Mihailović but wrong when convicting Alojzije Stepinac, and the other way around (Đureinović, 2018). Denouncing historical revisionism abroad does not imply criticising the same tendencies at home, but those persecuted in their own country are considered innocent victims of communism while those in the other country are seen as war criminals and fascists.

Discussing the 1946 trial

Already in December 2013 when Latas testified, Oliver Antić argued that there was no need to prolong the case any longer and demanded to hear the closing remarks (Viši sud, 2013a: 7). Antić's involvement changed the dynamics of the process. Although he formally represented the JVuO Association, he dominated the proceedings, constantly insisting on limiting the focus to the 1946 trial, complaining and trying to speed up the process by dismissing the usual steps of the judicial procedure. He continuously objected to the presence of Deputy Prosecutor Goran Rašić at the hearings, who observed the process because of public interest, without taking a position. The change of the judge presiding over the court council probably had the effect of speeding things up as well, so, as of early 2014, plaintiffs and the court ordered no additional materials or testimonies and the process ended in only a few proceedings.

Another novelty in the process was the audio recording of hearings, as opposed to the usual stenographic notes. This caused complications and trouble. The judge had to keep telling the audience to calm down to be able to properly record the hearing. All speakers had to start by saying their names into the microphone to be recognisable for the transcript, but the lawyers kept forgetting this and the judge had to remind them multiple times (Viši sud, 2015d: 5). Antić demanded they remain seated while speaking, which the judge first allowed but then banned, another reason for Antić to enter discussion with the judge, this time about whether standing while speaking is a must in this type of a court procedure (Viši sud, 2015a: 2).

In 2015, the case was coming to an end. The court announced introductory remarks where plaintiffs could revisit their rehabilitation requests, followed by evidential hearing and closing words. Wanting to rush the process, Antić and Živanović refrained from their introductory remarks (Viši sud, 2015d: 7). The evidential hearing revisited testimonies and all materials involved in the case, projected on a screen. The court also played the audio recordings of Mihailović's trial, although Antić protested and Živanović complained that he could not understand anything (Viši sud, 2015a: 7–9). The judges compromised by playing five minutes of each file and declaring they had heard all evidence. Pointing out people leaving the courtroom, Antić continued to complain (Viši sud, 2015a: 9).

Closing words

Having initially prepared 60 pages of text for his closing, Oliver Antić chose to present a third of it in the courtroom. His closing remarks lasted two hours, followed by Živanović's talk which lasted over an hour (Viši sud, 2015a: 11). Antić argued that the process had unnecessarily wasted time, deeming all "so-called testimonies" and "so-called expert opinions" legally irrelevant (Viši sud, 2015b: 3). He stressed that it only mattered whether a punishment was for political and ideological reasons, claiming that the court case could have ended after only two proceedings (Viši sud, 2015b: 4). He saw his involvement as a duty:

> The members of the Association of the Yugoslav Army in the Homeland 1941–1945 from Belgrade, came to me with the following words: "You returned our highest commander, King Petar II, to the country, give us back commander Draža too!". The youngest among them, who entered the 90s of his life, offered to bring the official invitation personally. That honourable elder was a young messenger for Đeneral Draža Mihailović and his biggest wish and honour was to complete his final task. To paraphrase, judges, the greatest jurist of the antiquity: their wishes I could not overhear, the suffering and victimisation of the members of the Yugoslav Army in the Homeland, led by Đeneral Mihailović, I could not ignore, nor the Golgotha of the Serbian people, the Serbian Orthodox Church and many honourable Yugoslavs, I could not forget, nor pass over.
>
> (Viši sud, 2015b: 3)

Calling Mihailović's execution a murder, Antić discussed the irregularities in the trial, such as no right to appeal and that Mihailović had been executed only 48 hours after his sentence. However, even though he drew attention to the trial, Antić engaged in the discussion on the positive qualities of Mihailović, just the same as the witnesses he had criticised previously. It was particularly important to emphasise the international reputation of Mihailović among the western Allies and United States presidents, namely Ronald Reagan. For Antić, Mihailović was "a convinced Yugoslav and a democrat" who managed to gather thousands of Croats, Muslims and Slovenes in the units under his command (Viši sud, 2015b: 20). Speaking poetically and pathetically about Mihailović and his destiny, Antić offered glorification rather than legal analysis. He emphasised that Mihailović loved French literature.

While "Draža" was an honourable man who loved his country and did not want to leave, even if that meant he had to die, "Broz", the communists, their regime and revolution were anti-Serb oriented, militarily and otherwise uneducated and only striving for power, according to Antić. He claimed that communists killed 50,000 Serbs and blamed the Partisans for starting the civil war and contributing to Serbs' suffering, a real high treason (Viši sud, 2015b: 12). He also blamed Tito for working together with the Ustasha, stating: "After all, who is attacking Draža Mihailović and the Yugoslav Army in the Homeland today? Those are precisely the old allies: Ustasha and neo-Ustasha organisations, Ballists and communists. No one else" (Viši sud, 2015b: 23).

Representation of communists as perpetrators involved the portrayal of the Serbian nation as their biggest victim. One of the most notable examples of victimhood that Antić underlined was the Serbian Orthodox Church, presented as more persecuted than other religious communities in socialist Yugoslavia (Viši sud, 2015b: 23). He finished by quoting Nikolaj Velimirović's depiction of Mihailović as falling for his nation like Kosovo knights and addressing the court to perform their holiest human and court duty and crown the case with a positive decision "on the great day of St. Basil of Ostrog, a saint and wonder maker" (Viši sud, 2015b: 30).

As opposed to Antić, Živanović focused on the legal issues, but not without evaluation of history. He addressed Mihailović as the commander of the first guerilla fighters in occupied Europe, persecuted by the occupation more than communists (Viši sud, 2015b: 31). Having survived four war years, Živanović explained, ideological and political opponents ended Mihailović's life after a farcical show trial. He pointed out that ten others convicted at the same trial lived in Great Britain or the United States and were never extradited, for Živanović a clear sign that no foreign government recognised the sentences (Viši sud, 2015b: 32).

Speaking about the trial, Živanović concentrated on two questions: whether ideological and political reasons for such a sentence existed during the trial and if the verdict was delivered because of those reasons. Interpreting Mihailović as the most dangerous enemy of the communists because of his reputation and power to bring all democratic forces together, Živanović claimed that the Communist Party had political and ideological reasons to deprive Mihailović of his life in the period of destruction of all opponents of the regime (Viši sud, 2015b: 33).

Addressing the second question, Živanović provided legal analysis and discussed the influence of the highest party structures on the trial, arguing that Mihailović's punishment was a retroactive implementation of the Law on Criminal Acts against People and State for the acts he had committed before the law was passed in 1945 (Viši sud, 2015b: 34). Moreover, the court did not apply a criminal proceeding law of the Kingdom of Yugoslavia, going through the interwar legislation and explaining in detail how the court had not respected it. Živanović's opinion is that the execution of Mihailović had been carried out in a way not characteristic of legal implementation of death sentences in a civilised country, because his burial site was secret. This confirms that "those who judged, nor those who carried out the punishment, had a clear conscience" (Viši sud, 2015b: 41). The communists acted like they had committed a crime (Viši sud, 2015b: 42).

Plaintiffs' claims summarised: the court decision

The court council decided positively about the request to rehabilitate Dragoljub Mihailović in May 2015. When judge Aleksandar Trešnjev read the decision in front of the full courtroom, the audience had to be calmed down for the explanation of the decision. Right away, the judge underlined that the subject of the process was not the rehabilitation of the Yugoslav Army in the Homeland as that had already been done by the 2004 changes to the Veteran Law. He continued: "The subject of this process was the legality of the trial conducted against Dragoljub Mihailović" (Viši sud, 2015c: 2).

The court decision summarised what had been said in the courtroom during the process, demonstrating that the judges did not seriously consider any facts that would speak against Mihailović's rehabilitation. The explanation of the decision to rehabilitate Mihailović listed several procedural irregularities of the 1946 trial, such as the obstruction of defence, predetermination of the trial and verdict by political actors, its rushing, forged evidence, the general atmosphere of the trial and the interference of the highest state authorities. Following the claims of the plaintiffs, their lawyers and witnesses, the court gave special attention and significance to the American airmen and their rejected attempt to travel to Yugoslavia to testify. According to the court, the process "was not fairly and correctly carried out by respecting all rights of the defendant, according to the standards of both contemporary law as well as the legal standards of that time" (Viši sud, 2015c: 2). Thus, Dragoljub Mihailović was to be rehabilitated as a victim of the trial manipulated for political and ideological reasons.

At the end of his speech, Trešnjev repeated that the court had not dealt with the facts and conclusions of whether Mihailović had been a war criminal, but their jurisdiction had been limited to evaluating if the trial had been fair and just, which it had not been. He concluded:

> We are convinced that the verdict had been made in advance, by politicians and not judges. Because of that, Dragoljub Mihailović stood no chance in this process. The procedure where the rights of the defence were so harshly violated, where there were no uncertainties, could surely not lead to a correctly determined fact situation and a right court decision based on it.
> (Viši sud, 2015c: 8)

The official court decision underlines this argument.

The official court decision document establishes a fact pattern on more than 30 pages before moving on to concisely justifying the positive decision (Viši sud u Beogradu, 2015). Here, too, the court considers only the claims, materials and witness testimonies the plaintiffs had delivered to the process. These are only the facts that speak in favour of Mihailović's rehabilitation.

While the decision paraphrases historians' testimonies and uses them as evidence, it only briefly describes the statement by Branko Latas, although it lasted several hearings. The description excludes his claims about the collaboration and crimes (Viši sud u Beogradu, 2015: 31). The decision explains that "the court did not especially consider the statement of historian Branko Latas, because the historian, as he himself said, never worked on the trial against Mihailović" (Viši sud u Beogradu, 2015: 33). On the other hand, the court seriously considers the statements by Bojan Dimitrijević and Kosta Nikolić, including their opinions about Mihailović's trial, even though they have not done any scholarly work on the trial either, but they are historians of the Second World War, as Latas was. While ruling out Latas for speaking about the Second World War and the role of the Partisans and Chetniks in it, the court decision provides details of Milton Friend's wartime reminiscence, including the remark about Mihailović wanting to stay with his people that Milton had heard from someone else (Viši sud u Beogradu, 2015: 32).

The repetition of the plaintiffs' and lawyers' claims drew emphasis on certain issues that could otherwise be deemed irrelevant for evaluating the 1946 trial and whether it had been rigged for political reasons. For instance, Trešnjev's closing speech and the court decision repeat that Mihailović was given strong alcoholic drinks while in custody, stressing the volume and how it increased. The court council concluded that the alcohol, together with the tempo of the trial of 25 days without a break, had influenced the quality of defence and made Mihailović confused while answering questions (Viši sud u Beogradu, 2015: 35). There was no evidence that Mihailović had been forced to drink and the guard had not testified about that. Furthermore, the lack of an official written report and the secrecy about Mihailović's execution and burial site, "not typical for legal implementation of death sentences in a civilised country", convinced the court undoubtedly that Mihailović had been killed for political and ideological reasons (Viši sud u Beogradu, 2015: 38). In this way, the judges ignored the wider context of the postwar trials, such as Nuremberg, where the executed defendants had also been deprived of marked graves. The court gave decisive relevance to the issue of an unmarked grave for the evaluation of the 1946 trial.

Notes

1 The interwar lawyer and politician who served as the prime minister in the Yugoslav government in exile during the Second World War. The 1946 Belgrade Process sentenced him to 20 years in absence, depriving him of rights, citizenship and property, while he was in London where he stayed for the rest of his life.
2 Helsinki Committee for Human Rights in Serbia, Women in Black, the Independent Association of Journalists of Vojvodina and the Society for the Truth about the People's Liberation War (*Društvo za istinu o NOB-u*). The Association of Antifascists emerged from SUBNOR because of SUBNOR's support for Slobodan Milošević in the 1990s and the later inclusion of the 1990s veterans in the association. They are significantly smaller, encompassing only a few hundred members as opposed to SUBNOR that has thousands. It is not an official veteran association but includes the Partisan veterans who were opposed to Milošević's regime. The two organisations do not cooperate, seeing the reasons for the split as irreconcilable.
3 As opposed to hegemonic discourses, there is cooperation between revisionist historians from Serbia and Croatia, based on anti-communist consensus, even though they are

promoting incompatible nationalist narratives. The most notable example is the cooperation between Bojan Dimitrijević and Zlatko Hasanbegović, historian, politician and the minister of culture in Croatia in 2016, who is involved in, among other things, the Bleiburg commemorations. Their cooperation involved subsidies for three books by Dimitrijević published in Croatia, which both Serbian and Croatian media attacked. In addition to that, Cvetković and Nikolić took part in conferences at the Croatian Institute of History. The biggest controversy involved a photo of a celebration of a publication in 2013, posted online by Despot Infinitus publisher from Croatia, showing Dimitrijević wearing a shirt with the symbol of the Prinz Eugen division.

Bibliography

Beta (2010) "Slučaj Mihailović: Sud odbio svedočenje partizana". Available at: www.novosti.rs/vesti/naslovna/aktuelno.69.html:310770-Slucaj-MihailovicSud-odbio-svedocenje-partizana (accessed 22 October 2018).

Blic (2015) "Princ Aleksandar pozdravio rehabilitaciju Draže Mihailovića". Available at: www.blic.rs/vesti/drustvo/princ-aleksandar-pozdravio-rehabilitaciju-draze-mihailovica/gn1ferd (accessed 31 October 2018).

Drašković, V. (2016) Interview by author.

Dulić, T. (2012) "Sentenced 'for Ideological and Political Reasons'?: The Rehabilitation of Dragoljub Draža Mihailović and Social Memory in Serbia", *Sociologija* LIV(4): 625–648.

Đureinović, J. (2018) "(Trans)national Memories of the Common Past in the Post-Yugoslav Space". In Berger, S. and Tekin, C. (eds), *History and Belonging: Representations of the Past in Contemporary European Politics*. New York: Berghahn Books.

Glavonjić, Z. and Mihajlović, B. (2015) "Rehabilitacija Draže: Revizija istorije, slavlje i ogorčenje". Available at: www.slobodnaevropa.org/a/sud-rehabilitovao-drazu-mihailovica/27016033.html (accessed 31 October 2018).

Krivična prijava (2013). Available at: www.helsinki.org.rs/serbian/doc/krivicna%20prijava%20hos%20sept_17_2013.pdf.

Latas, B. (1999) *Saradnja četnika Draže Mihailovića sa okupatorima i ustašama: 1941–1945*. Belgrade: Društvo za istinu o antifašističkoj narodnooslobodilačkoj borbi 1941–1945.

Mackenzie, W. (2002) *The Secret History of SOE: Special Operations Executive 1940–1945*. London: St Ermin's Press.

Milekić, S., Ristić, M. and Džidić, D. (2015) "Hrvatska predsednica kritikuje rehabilitaciju Draže". Available at: www.balkaninsight.com/en/article/hrvatska-predsednica-kritikuje-rehabilitaciju-dra%C5%BEe/1431/145 (accessed 31 October 2018).

Nezavisne novine (2012) "BiH razmatra povlačenje ambasadora zbog Draže Mihailovića?" Available at: www.nezavisne.com/novosti/bih/BiH-razmatra-povlacenje-ambasadora-zbog-Draze-Mihailovica/134760 (accessed 31 October 2018).

Nova srpska politička misao (2011) "Srpski liberalni savet: Ko je pravi pokretač procesa rehabilitacije generala Mihailovića". Available at: www.nspm.rs/hronika/srpski-liberalni-savet-ko-je-pravi-pokretac-procesa-rehabilitacije-generala-mihailovica.html?alphabet=l (accessed 29 October 2018).

Pejčić, I. (2009) "Rehabilitovan advokat Slobodan Subotić". Available at: www.danas.rs/hronika/rehabilitovan-advokat-slobodan-subotic/ (accessed 28 October 2018).

Spasojević, V. C. (2012) "Rehabilitacija Draže: Hrvati spremni samo za mržnju". Available at: www.novosti.rs/vesti/naslovna/drustvo/aktuelno.290.html:373403-Rehabilitacija-Draze-Hrvati-spremni-samo-za-mrznju (accessed 31 October 2018).

Večernje novosti (2012) "HDZ: Povući ambasadora iz Srbije ako rehabilituju Dražu". Available at: www.novosti.rs/vesti/planeta.300.html:372809-HDZ-Povuci-ambasadora-iz-Srbije-ako-rehabilituju-Drazu (accessed 31 October 2018).
Viši sud (2010a) Zapisnik sastavljen pred Višim sudom u Beogradu dana 09.12.2010. godine. Belgrade: Viši sud u Beogradu.
Viši sud (2010b) Zapisnik sastavljen pred Višim sudom u Beogradu dana 29.10.2010. godine. Belgrade: Viši sud u Beogradu.
Viši sud (2011) Zapisnik sastavljen pred Višim sudom u Beogradu dana 30.11.2011. godine. Belgrade: Viši sud u Beogradu.
Viši sud (2012a) Zapisnik sastavljen pred Višim sudom u Beogradu dana 11.05.2012. godine. Belgrade: Viši sud u Beogradu.
Viši sud (2012b) Zapisnik sastavljen pred Višim sudom u Beogradu dana 20.06.2012. godine. Belgrade: Viši sud u Beogradu.
Viši sud (2012c) Zapisnik sastavljen pred Višim sudom u Beogradu dana 22.11.2012. godine. Belgrade: Viši sud u Beogradu.
Viši sud (2012d) Zapisnik sastavljen pred Višim sudom u Beogradu dana 23.03.2012. godine. Belgrade: Viši sud u Beogradu.
Viši sud (2012e) Zapisnik sastavljen pred Višim sudom u Beogradu dana 27.01.2012. godine. Belgrade: Viši sud u Beogradu.
Viši sud (2013a) Zapisnik sastavljen pred Višim sudom u Beogradu dana 24.12.2013. godine. Belgrade: Viši sud u Beogradu.
Viši sud (2013b) Zapisnik sastavljen pred Višim sudom u Beogradu dana 25.01.2013. godine. Belgrade: Viši sud u Beogradu.
Viši sud (2014) Zapisnik sastavljen pred Višim sudom u Beogradu dana 04.04.2014. godine. Belgrade: Viši sud u Beogradu.
Viši sud (2015a) Transkript audio zapisa sa ročišta povodom zahteva predlagača za rehabilitaciju Dragoljuba Mihailovića održanog dana 03. aprila 2015. godine. Belgrade: Viši sud u Beogradu.
Viši sud (2015b) Transkript audio zapisa sa ročišta povodom zahteva predlagača za rehabilitaciju Dragoljuba Mihailovića održanog dana 12. maja 2015. godine. Belgrade: Viši sud u Beogradu.
Viši sud (2015c) Transkript audio zapisa sa ročišta povodom zahteva predlagača za rehabilitaciju Dragoljuba Mihailovića održanog dana 14. maja 2015. godine. Belgrade: Viši sud u Beogradu.
Viši sud (2015d) Transkript audio zapisa sa ročišta povodom zahteva predlagača za rehabilitaciju Dragoljuba Mihailovića održanog dana 27. februara 2015. godine. Belgrade: Viši sud u Beogradu.
Viši sud u Beogradu (2015) "Rešenje Reh. 69/10". Available at: www.bg.vi.sud.rs/tekst/125/rehabilitovani-i-odbijeni-zahtevi.php (accessed 22 October 2018).
Živanović, Z. (2016) Interview by author.

Conclusion

As Tony Judt observes, the "mismemory" of communism is contributing to a mismemory of anti-communism (Judt, 2000: 309). This is remarkably fitting to the context of memory politics in post-Milošević Serbia, where it is impossible to disentangle the memory of the Second World War from the memory of the vanished socialist state of Yugoslavia. The long processes of unmaking of the Partisan myth and making of the Chetnik one have developed in parallel and they endure in the state of inseparable entwinement. Delegitimisation of the Partisans and Yugoslav state socialism is, however, not a by-product of the positive recasting of the Chetniks. It is rather the initial aim of the anti-communist memory work. The image of the Chetniks as an exclusively positive historical reference and their statues of ideal ancestors for the Serbian nation-state represents the epilogue of the decades-long efforts. These efforts are driven by anti-communism, often in combination with Serbian ethnonationalism.

The history of memory of the Second World War and its aftermath in Serbia is a story of continuities regarding narratives, actors and practices and ruptures that have echoed through decades. Changing understandings of collaboration, resistance and victimhood under communism underlie these processes. The diachronic approach enables us to observe these continuities and ruptures and trace the emergence and disappearance of practices and actors and long-term transformations in memory politics and memory cultures. A combination of a diachronic perspective with a profound analysis of the contemporary context fosters comprehension of the politicalness and historicity of memory cultures.

The historically contextualised in-depth analysis does not suffice, if a memory culture is observed in isolation, because memory politics is always intertwined with other countries of the post-Yugoslav space and with the broader international context, in the case of Serbia. Memory politics reflects positionalities in the international political setting. The attitudes towards international actors, like the European Union or Russia, inform politics of memory of a nation-state, as they inform the images of the Chetniks within Serbia. In the case of the former adversaries in an armed conflict, many of which exist in the post-Yugoslav space, mnemonic conflicts take multiple forms – from mere incompatibleness and mutual exclusivity of dominant narratives, through gloating over the neighbour's misdeeds and to competitive victimhood. At the same time, the hegemonic historical narratives in both contexts can be very similar. For Serbia, Croatia is the most important mnemonic adversary. The dynamics between the two countries involve the mutual exclusivity of otherwise analogous narratives about the Second World War and Yugoslavia, gloating about the neighbour's perpetrating nature in the past and historical revisionism in the present and competitive victimhood.

National fragmentation of the Yugoslav memory culture is not a new phenomenon and it was there in state socialism. It stretches from the simple observation of the past through the lens of a nation to the competitive narratives and memory cultures, as they developed between Croatia and Serbia. Nationalisation of memory involves othering of socialist Yugoslavia and de-Yugoslavisation of the Partisans as Serbian. Serbian people become antifascist, regardless if they chose to join the Partisans or the Chetniks, and the Serbian nation is depicted to have stood on the right side throughout the history. The narrative of national reconciliation solves the issue of internal conflicts and harmonises the opposing sides of civil war within the Second World War, framing everyone as having ultimately fought for a national cause. This narrative featured prominently in Croatia too, especially in the 1990s, where the victorious and the defeated to be reconciled were the Partisans and Ustasha.

Historical victimisation of the Serbian nation occupies the central position in the nationalist remoulding of the past that came to the foreground in the 1980s, pointing fingers at and ascribing collective guilt to perpetrators and others who could be blamed for the nation's tragic destiny. The genocide against Serbs in the Independent State of Croatia logically became a highly relevant theme but so did the decades of living in Yugoslav state socialism. The idea of conspiracy against the Serbian nation, at both international and intra-Yugoslav level, underscored the discussions on the Second World War and socialist Yugoslavia.

The 1980s debates spilled over to the ambivalent era of Slobodan Milošević's rule in what was left of Yugoslavia. While waging wars in other Yugoslav republics and as these wars, inflation, sanctions and despair shook rump Yugoslavia, the regime promoted itself as successors of Yugoslavia and keepers of the legacies of the People's Liberation War. The Partisans' struggle was depoliticised, ethnicised as Serbian and observed through the dichotomic lens of the victimhood and heroism of the Serbian nation. In this way, the regime could represent itself as the bulwark against fascism, fascism being a metaphor for the secessionist claims of other Yugoslav people, particularly Croatia. The state did not endorse the Chetnik revival and commemorations that emerged in the early 1990s among the political opposition, but they were tolerated.

As this book demonstrates, there are numerous continuities not only regarding patterns of historical interpretation, but also concerning the actors promoting them. Involving many participants of the 1980s debates, the oppositional political parties that formed in 1990 in Serbia had anti-communism as a common and one of the most important characteristics, striving for dealing with the communist past and fighting communism in the present. As they replaced Milošević in 2000, the struggle against communism became the central objective of state-sanctioned memory politics, equally informing state policies in other spheres.

The commemorations of the Chetniks emerged in the early 1990s, with the SPO and Vuk Drašković as the main carriers of their memory which obtained massive popular support. This is when the first divisions in the interpretations of the Chetnik movement among their numerous adherents manifested themselves, as the wars in Croatia and Bosnia were gaining momentum. The dividing line was the understanding of the nature of the Chetnik movement and the attitude towards the wars of the Yugoslav succession and Serbian participation in them. In the post-Milošević period, the position towards the European Union surfaced as an additional demarcation line, while the divisions and the divided among the Chetnik memory community remain the same. The SPO and their affiliates promote the image of the Chetniks as democratic, Yugoslav and oriented towards the West. On the other hand, the SRS led by Vojislav Šešelj, as the most prominent actors of the opposing side of the Chetnik interpretation, perceive the Chetnik tradition as striving for uniting all Serbs in one ethnically homogeneous state, thus, justifying wars to achieve it.

The Serbian Orthodox Church has been a key agent of anti-communist memory politics since the early 1990s. Religious memorial services and requiems for victims of communism, including the Chetniks and other defeated forces of the Second World War, surfaced as a common commemorative practice. The clergy blesses every memorial or plaque honouring victims of communism. In Serbia, there are two churches dedicated specifically to national reconciliation and victims of communist terror respectively.

Whereas the multitude of continuities is evident, regarding narratives, actors and practices at all levels of memory work, the overthrow of Slobodan Milošević in 2000 also embodies a turn concerning the memory of the Second World War and socialist Yugoslavia. With the political opposition to Milošević coming to power, anti-communism became the state policy and foundation of the institutional level of memory work. The state actors dedicated enormous efforts into the removal of references to socialist Yugoslavia from the public sphere, rehabilitation of the Partisans' wartime opponents and recognition of those now officially considered victims of communism. These efforts were the result of the very strong anti-communist consensus and the opposition was marginal. At the same time, the state institutions did not seriously tackle the wars of the 1990s and their legacies.

The Yugoslav Army in the Homeland constitutes the central theme of the institutional level of memory work in Serbia that is based on the parallel revision of socialist Yugoslavia and the Second World War as its main legitimacy source. Within the hegemonic discourses, the Chetniks are both an antifascist movement equal to the Partisans and victims of communism, with their leader Dragoljub Mihailović symbolising the victimisation of the entire movement by Yugoslav communists. In addition to being considered unconditionally antifascist throughout the whole duration of the Second World War, which is historically inaccurate, the Chetniks are ascribed the epithet they never used to define themselves.

168 Conclusion

The equalisation of the two movements is an aspect of the wider politics of national reconciliation, that, in the case of the Second World War, should finally end the civil war that the Partisans started on 7 July 1941. What is more, a non-communist and anti-communist resistance movement appears in Serbian history to replace the communist-led Partisans as a positive historical reference.

The equalisation of the two movements is apparent. Negative revaluation of the Partisan movement has been an inseparable dimension of the positive recasting of the Chetniks for decades. The emphasis on the victimisation of the Chetnik movement through the example of Mihailović's 1946 trial and execution by the Yugoslav authorities directly implies the Partisans and the Yugoslav regime as perpetrators. Moreover, Dragoljub Mihailović not only symbolises the suffering of the Chetnik movement, but also the victimhood of the Serbian nation under communism. Accordingly, all state efforts and public representations of the past focusing on victims of communism revolve around the Chetnik movement. As I show throughout this book, this is the case with state-instituted initiatives such as investigative commissions, with museal and media representations and with the anti-communist memory work coming from below. The Partisans, on the other hand, are either deemed unsuitable for the current political order or occasionally evoked; however, only in a nationalised and de-politicised variant, their victory is celebrated as a Serbian victory in the Second World War.

The equivocal nature of the Chetnik movement makes them a convenient and uncomplicated object of reinterpretation founded on selectivity. While the Chetnik ambiguity enables positive reinterpretation and construction as an exclusively favourable historical reference, it also generates diverse interpretations. As the divisions among the Chetnik adherents in the 1990s illuminate, there is more than one image of the Chetnik movement. By the same token, each actor claiming continuity with Mihailović's Chetniks or a proponent of their rehabilitation has their own "Draža".

For some actors, Dragoljub Mihailović remains a Yugoslav-oriented democrat who admired Western democracies and the European Union, even before it existed. For others, Moljević's *Homogeneous Serbia* is the most attractive dimension of the Chetniks, understanding them as anti-Western and striving for the unification of all Serbs within the state borders of Greater Serbia. The monarchist identity of Mihailović's forces is another important point, for some merely an argument in favour of Chetniks' legitimacy as "the King's army", as opposed to the illegitimate Partisans, while for others an actual political agenda of today. Finally, what most of these actors can agree upon and continuously emphasise is that Mihailović loved French literature, a characteristic for some reason so important that it entered post-2000 history textbooks and courtroom debates.

The multiple sides of the Chetnik memory intertwine with positionalities concerning both internal and external issues. Already in the early 1990s, the attitude towards the wars of the Yugoslav dissolution and the Serbian participation in it created a chasm between the political parties affiliated with the Chetnik tradition. As new groups and political parties emerged later, they positioned themselves on one side. In addition to this, the positionality towards the European Union and Russian Federations, important external memory agents, is another line that divides the contemporary Chetnik adherents and the Chetnik images they cherish. For this reason, men wearing Chetnik symbols fighting in eastern Ukraine and demands for a memorial to Mihailović in Brussels are public manifestations of the same but divided memory community.

Diverse images of the Chetniks illuminate that mnemonic hegemony does not imply homogeneity. The all-encompassing hegemony and constant public presence of revisionist narratives goes hand in hand with multiple actors, divisions, oppositions, negotiations and challenges that all stretch across multiple levels of memory work. Behind all that, however, is the anti-communist consensus of the post-Milošević political elites. The Chetnik revaluation is not a matter of right-wing groups and parties, as Chetnik followers are often simplified. The SPO, SRS or the black-clad members of the Ravna Gora Movement did not politically and legally rehabilitate the Chetniks alone. The Chetniks are central to Serbia's state-sponsored memory politics after Milošević. The DS and parties affiliated with it either generated or strongly supported most initiatives since 2000, too. The consensus is evident in the clear majority voting in favour of legislation that regulates historical interpretation and the marginal criticism and opposition in state institutions.

The centrality and primacy of the Chetniks within memory politics corresponds to the superficial nature of the state interest for all other groups considered victims of communism, that usually operates at the rhetorical level. To illustrate once again, the first fact-finding commission established in Serbia was dedicated only to the search for the grave of Dragoljub Mihailović and it was the government that established it. On the other hand, the investigative commission dedicated to the postwar executions in general emerged afterwards, only because of lobbying by historians and a daily newspaper. The commissions support the argument of Serbia's state-sanctioned memory politics as both official and informal, frivolous and shaped by a very strong personal dimension. They also demonstrate how state endeavours of investigative commissions can end unsuccessfully in failure and embarrassing public debacles.

That the Chetniks matter the most among the victims of communism does not mean that reparations to them are any less symbolic. The objective of state-sanctioned memory politics is their political and symbolic rehabilitation as a historical reference and the similarly rhetorical condemnation of the Partisans and socialist Yugoslavia. This does not necessarily include the actual care for the still living Chetnik veterans or the families of those who had died, even when memory politics is narrated as having to do with human rights and transitional justice. In a like manner, no one really cares about Dragoljub Mihailović nor his family. Therefore, the Veteran Law revision, as problematic as it was, has not resulted in pensions or any benefits for the Chetnik veterans and their families, remaining merely the legal confirmation of the positive image of the Chetnik movement and their equalisation with the Partisans.

In a similar manner, the rehabilitation process of Dragoljub Mihailović did not lead to any form of compensation for his descendants nor did it resonate in

170 *Conclusion*

the memoryscape in the form of an official memorial or the establishment of regular commemorative practices honouring him and the Chetniks. The significance of the process is bringing the image of Mihailović as an innocent victim of communism to the forefront through the tremendous media attention the court case generated and making this image official by the positive outcome of the court case. While the court continuously emphasised that the case was only about the question of whether Mihailović had been sentenced and executed because of political and ideological reasons, the Second World War was the central theme of court hearings, including the discussion of the nature of the Chetnik movement, their collaboration and crimes. The focus on the wartime context transcended the nature of the process as only righting the past wrongs, evaluating the Second World War and its most prominent actors.

The failure of the state efforts and the lack of memorials, official holidays and commemorations honouring the defeated side of the Second World War and victims of communism caused frustration among the actors who had expected the political changes of 2000 to include serious efforts into dealing with the communist past. This led to them organising themselves as memory entrepreneurs at the social level of memory work. As other levels of memory work, this sphere is also shaped by numerous continuities with the 1980s and the 1990s, while the narratives they promote reflect the hegemonic discourses. The centrality of Dragoljub Mihailović and the Chetniks is equally evident in these memory communities and anti-communism serves the unifying factor.

While the anti-communist memory community shares the discourses that are hegemonic in Serbian society, it involves a constant struggle for recognition. Specific memorials honouring victims of communism and the Chetnik movement respectively, supported by the state and located in Belgrade, represent the form of acknowledgement these groups strive for, even though their numerous efforts have not resulted in the approval of the city and state authorities so far. The mechanism of judicial rehabilitation is one of the forms of seeking recognition, as the legislation provided the space for the actors from below to get involved as plaintiffs and have their claims heard in the courtroom and in public. The law, thus, is not only the mirror and instrument of memory politics, but also a contact zone between the institutional and other levels of memory work in Serbian society.

The multifaceted approach of this book has implications that go beyond the Serbian case. This study empirically explores the phenomenon of memory politics in a national context, using sources of various provenances. It combines the diachronic perspective that illuminates the historical variability of memory, while pointing out numerous continuities of historical narratives, practices and actors, with in-depth analysis of contemporary memory politics in post-Milošević Serbia. In doing so, the aim of this book is to demonstrate the multiple levels of memory work and their interactions, as well as heterogeneity that underlies the mnemonic hegemony. As such, the book centres on the national frameworks of memory, demonstrating that the nation-state retains its relevance in the context of the transcultural and arguing for the approach to particular memory cultures that fragment the global perspective. One way to do this is to take an empirical approach to memory cultures and politics, as this book does. While it represents an actor-oriented, in-depth analysis of the Serbian case, the book brings it into communication with the wider context by demonstrating that the mnemonic tendencies in Serbia do not exist in isolation. However, a wider approach that would seriously tackle the correlations, comparisons and parallels outside of the Serbian context remains beyond the scope of this study.

Bibliography

Judt, T. (2000) "The Past Is Another Country: Myth and Memory in Postwar Europe". In Deák, I., Gross, J. and Judt, T. (eds), *The Politics of Retribution in Europe: World War II and Its Aftermath*. New Jersey: Princeton University Press, pp. 293–324.

Index

1944 Commission 87, 92, 93, 95, 104, 120, 143

Ada Ciganlija 100, 101, 103–104, 120
Antić, Čedomir 73, 105n2
Antić, Oliver 1, 147, 156, 158–160
anti-communism 6–7, 14, 56, 66–67, 72, 96, 111
anti-communist consensus 6–7, 75, 122, 167, 169
anti-communist memory community 111–112, 142, 151; of Belgrade 121
antifascism 14, 72; as a European value 74; narrative of Serbian 2, 48
Association of Political Prisoners and Victims of the Communist Regime 110, 143, 149; see also Association of Victims
Association of Victims 120, 121, 127, 144, 151

Bosnia 57, 74, 126, 147, 157, 167; and Herzegovina 39, 73, 108, 157

Čavoški, Kosta 56, 122, 144
Čiča 63n7
civil society 67, 74
civil war 15, 48, 56, 79, 140–142; trauma of 113; within the Serbian nation 67, 70, 116
collaborators 6, 41–42, 68; as innocent victims of communism 81; rehabilitation of 130; settling accounts with 79; see also quislings
commemorative practice 38, 109, 122; from below 115
Communist Party of Yugoslavia 35, 45, 70, 140, 160
communist terror 13, 61, 80; victims of 2, 6, 59, 68, 125, 167

Croatia 57, 74, 108, 131; Serbia and 13, 157
Cvetković, Srđan 80–81, 90, 93–96, 99, 101, 121, 125

dark night 56, 61, 70, 76
Day of the Liberation 1–2, 54, 69; see also Liberation Day
decommunisation 10–11
Democratic Party 52, 56, 58, 71, 75, 169
descendants 98, 110–111, 113–114, 142
Dimitrijević, Bojan 87–88, 110, 122, 153–154
Drašković, Vuk 56, 58–61, 73, 150
Draža 59, 63n7, 73
DS see Democratic Party
Đuretić, Veselin 46–48, 152, 154–155

ethnicisation 2, 6–7, 55; see also national fragmentation of memory; nationalisation
ethnonationalism 55, 165
European Union 74–75
Europeanisation 74
exhumation 91, 94, 101–103; Board for 95–96; test 94, 97, 102

fact-finding commissions 11, 87, 104

Greater Serbia 3, 56, 168, 171; see also Homogeneous Serbia

hegemony see mnemonic hegemony
historians 30, 47–48, 87; anti-communist 90, 144; critical 155–156; as expert witnesses 139, 152–154; of memory 20; revisionist 67, 74, 93
history of memory 20, 165; see also mnemohistory

holidays: official 67, 69; public 110, 116
Homen, Slobodan 89–90, 100, 102, 121
Homogeneous Serbia 3, 14, 61, 168

Independent State of Croatia 59, 131, 139, 166

Kalabić, Nikola 138–140
Karađorđević 58: Aleksandar 122, 123, 150; family 109, 111, 123; Jelisaveta 110, 121, 123
Koštunica, Vojislav 58, 62, 123

Lazarev, Gojko 113–114, 125, 140–143
League of Communists 52
Liberation Day 6, 125
Lončar, Bogdan and Milenko Braković 39, 116, 140–142

Marković, Slobodan 80, 90, 152–153
memorial church 62, 113, 125
memorials: Chetnik 109, 115, 118–119, 125, 126; neglected 42; Partisan 36, 38, 40; to victims of communism 110, 124
memory community 1, 109; anti-communist 111–112, 121, 122, 124, 142–144, 148; Chetnik 59, 118, 167
memory culture 22–23, 53, 165–166; hegemony of the Yugoslav 112, 115, 121; nation-centred 9; post-Yugoslav 76; Yugoslav 36, 40, 41, 33, 57; *see also* memory subcultures
memory entrepreneurs 25, 43, 93, 104; *see also* mnemonic agents
memory laws 10–11
memory milieu 15, 23, 24
memory politics 6, 8, 22, 24, 53, 100; anti-communist 66; dominant 67, 110; dynamics of 24; revisionist 74; Serbia's 68; state-sanctioned 8, 112, 120, 130; state-sponsored 22–23, 169; Yugoslav 39
memory subcultures 23–24
memory work 25; anti-communist 7, 12, 109–112, 119, 122, 165; hierarchy 26, 66, 120; at the institutional level 25, 36, 68, 70, 92, 167; levels of 10, 59, 95, 99, 148; multi-level 12, 25; social level of 25, 170
Mihailović Commission 87–90, 100, 103
Milošević 7, 36, 52–53, 111, 166; era 6, 52; opposition to 55–56, 123; overthrow of 66, 69, 70, 167; post- 6, 66, 70, 75, 82, 90; regime 53–54, 116

mnemohistory 21
mnemonic agents 25, 48, 88, 104; pro-Chetnik 73
mnemonic hegemony 20, 25, 66, 169–170
musealisation 77–78

narodni neprijatelj 5, 35
national fragmentation 9, 157, 166
national reconciliation 7, 13, 56, 67, 70, 117, 131, 140, 166
nationalisation 8, 65, 79, 158, 166
nationalism 6, 14, 45; Serbian 46, 71
Nedić, Milan 3, 5, 43, 138, 139
NGOs 111, 155, 156
Nikolić, Kosta 88, 122, 140, 152–153, 155, 156
Nikolić, Tomislav 72, 123, 151

official calendar 110

parliament 69–70, 72, 135
parliamentary debate 70, 75
Partisan myth 36, 41–42, 53
Patriarch Irinej 117, 123, 150
Pavlović, Momčilo 90–92, 95, 101
people's enemy 5, 35, 41–42, 114, 138, 143
People's Liberation War 6, 36–37, 45, 53–54, 68, 116; veterans 131–133
Petranović, Branko 46–47
Pogledi 57, 79, 129
political and ideological reasons 130, 138, 159, 161, 170
Popular Memory Group 22, 24
postsocialism 8–9, 11

Ravna Gora 57–58, 59, 60, 62, 118, 149; gathering 59, 110, 117–118, 123; memorial 61, 115; Movement 62, 74–75, 108–109, 117–118, 126, 147, 149
Ravna Gora 75–76
Red Army 9, 37, 39–40, 73
Rehabilitation Law 10–11, 130, 135–136, 137, 156
religious memorial service 58, 59, 110, 122; *see also* requiem
Republika Srpska 74, 108, 117, 119, 157
requiem 116, 117
resistance 3–4, 37, 43, 78, 155; Chetniks as 72, 75, 134
retribution 5, 7, 130
Romanovs 74–75, 117–118
Russia 1–2, 11, 74–75, 118, 165

Samardžić, Miloslav 57, 115, 129
Serbian Orthodox Church 58, 60, 111, 116, 118, 122, 124, 150, 167
Serbian Progressive Party *see* SNS
Serbian Radical Party *see* SRS
Serbian Renewal Movement *see* SPO
Šešelj, Vojislav 56, 73, 129, 167
SNS 8, 65, 82n3
SPO 56–57, 59–62, 71, 73–74, 117–118, 119
SRS 61, 72, 73
State Commission for Investigation of Circumstances of the Execution of Dragoljub Mihailović 87; *see also* Mihailović Commission
State Commission for Secret Graves 80, 87, 90, 143; *see also* 1944 Commission
street names 67, 69, 110
SUBNOR 37–38, 41, 54, 134

Tadić, Boris 70
Tito 4, 40, 41, 43, 55, 78; biography of 44; death of 44; Josip Broz 37, 55, 129
totalitarian legacies 80, 141
totalitarian past 120
transitional justice 12; faux 11; human rights discourses and 67, 169; pseudo 10–11
transnationalism in reverse 9

U ime naroda: association 110, 121; exhibition 80–81, 87, 121, 123
U ime naroda – za slobodnu Srbiju 100
Užice 4, 77–78, 115

Veteran Law 11, 70, 72, 131–133
victimhood 7, 36, 166; competitive 166; discourses of 114, 130, 142; Mihailović's 148; Serbian 45, 46; under communism 11, 58, 68, 79, 109, 113, 125, 165
violence 93, 136; communist 67; Partisans' 79; retributive 5, 35; revolutionary 68; victims of political persecution and 140, 142; victims of political repression and 10, 135
Vučić, Aleksandar 65, 72

Yugoslavia: dissolution of 61, 157; interwar 47; Kingdom of 2, 69, 70, 124; Milošević's 55; rump 52, 54, 62, 66; Second World War in 2, 37, 48, 72, 132; socialist 4, 6, 35, 36, 54–54, 66, 68, 112, 114, 166–167

Živanović, Zoran 144, 148–149, 160

Printed in the United States
By Bookmasters